Entropy and
Sustainable Growth

Other World Scientific titles by the author

The Entropy Crisis
ISBN: 978-981-277-968-7

New Superconductors: From Granular to High T_C
ISBN: 978-981-02-3089-0

Coherence in High Temperature Superconductors
ISBN: 978-981-02-2650-3

Entropy and Sustainable Growth

Guy Deutscher

Tel Aviv University, Israel

 **World Scientific**

NEW JERSEY · LONDON · SINGAPORE · BEIJING · SHANGHAI · HONG KONG · TAIPEI · CHENNAI

Published by

World Scientific Publishing Co. Pte. Ltd.

5 Toh Tuck Link, Singapore 596224

USA office: 27 Warren Street, Suite 401-402, Hackensack, NJ 07601

UK office: 57 Shelton Street, Covent Garden, London WC2H 9HE

British Library Cataloguing-in-Publication Data
A catalogue record for this book is available from the British Library.

ENTROPY AND SUSTAINABLE GROWTH

ISBN 978-981-3237-76-6

For any available supplementary material, please visit
http://www.worldscientific.com/worldscibooks/10.1142/10927#t=suppl

Printed in Singapore

In tribute to Roger Maynard

This book has been written in memory of Roger Maynard, to whom I was attached by a longstanding friendship that grew and deepened through our meetings and our discussions regarding a wide range of physics issues. His broad-minded vision and generosity were reflected in his sincere interest in the works of his colleagues. There were many of us who enjoyed these exchanges with Roger Maynard and we remain forever grateful. His election as Chairman of the French Society of Physics was an appointment he was so worthy of. His untimely demise has been a great loss for all of us.

In recent years, our discussions focused on the disturbing increase of entropy in the atmosphere, a topic I had raised in my book *"The Entropy Crisis"*. The outcome of these discussions was to have been a text written together, but destiny decided otherwise.

This book is deeply inspired by our discussions. It also expresses the desire for dissemination to a wider audience, a dissemination to which Roger Maynard attached great importance.

This book is also an opportunity for me to thank Rosette Maynard for her hospitality throughout these years during which our discussions were most often held in a family setting.

This text has benefited from Aline Deutscher's devoted reading, who also wished to take part in this tribute by contributing the painting "Finitude" that can be seen on the cover.

INTRODUCTION

Throughout history, men have left their countries, their birthplace in order to go elsewhere. Some were coerced such as the blacks who were forced into slavery, while others did it of their own free will. It could be due to massive migrations such as those we witness today, in the case of migrants who leave Africa and the Middle-East aiming to reach Europe, or it could have been based on personal decisions. Classical literature offers us well-known examples such as Ulysses' long mythical journey from which he returns or Abraham's one-way journey to the Promised Land of Canaan. We could also refer to the French author, Marcel Pagnol, and his famous hero, Marius, who is deeply attracted by the sea, a temptation to which he surrenders despite his great love for Fanny and his strong ties to the French port of Marseille. And at present, we are obviously being exposed to the Odyssey of Space which seems to be humanity's great challenge as was Colombus' voyage in the 15th century.

But why leave when it's so cozy at home? Abraham had a nice life in Haran near his father, Tera. Together with his cousin, Lot, they had established a wealthy well-to-do family. Why leave all that? What was Abraham to expect from this unknown Land of Canaan? What assurances had he been given to encourage him to obey the command to go there? He was promised the opportunity to be able to multiply, and Canaan was to serve as a space to be filled with his progeny. Today, we would say that the vision of growth was predominant, a growth which could not be further extended within a given site. Abraham decided to leave in order to escape from the *Finitude* of Haran where, despite everything, he had led a comfortable life. Likewise, Marius felt trapped in his port of Marseille. He needed to explore islands accessible only through deep sea navigation — a journey that, he hoped, would fulfill and empower him. Along the same lines, the heroine of the film, "Brooklyn", decides to leave Ireland when she realizes the narrow mindedness of her social environment. She is determined to put down roots for a new life in America although she can have a good life in her home city. The attraction of open space where everything is yet to be built, is stronger than her interest in a sweet home where everything awaits her.

Referring to a concrete and updated example, one might ask why do societies invest huge amounts of money in projects such as a space station or the conquest of Mars, while on Earth there is so much suffering to be resolved. And why are there so many volunteers wishing to reach Mars, which is not, we can suppose, Paradise? Should these collective and individual decisions be considered as a reaction to a striking malaise, now that we are aware of the finitude of our planet, of growth perspectives that appear to be decreasing?

From ancient times till today, whether in Greek mythology, Judeo-Christian civilization or exploits of modern times, the theme of incompatibility between finitude and growth is always present.

This book will indeed try to enlighten the meaning of finitude and growth nowadays. Their incompatibility has been clearly stated by Malthus. In his book, published in 1798, *Essay on the Principle of Population*, finitude is associated with arable lands and inevitably one day it will put an end to the growth of agricultural production and thus to the growth of population. According to Malthus, human beings are especially motivated by growth, and its end may well lead to an acute crisis. This theme has recently been rediscussed and developed by the Club of Rome as detailed in the next chapter.

Growth is indeed a fundamental characteristic of the living, either on a collective or on an individual scale. Thus, it is not surprising that announcing its inevitable end, is accepted by our societies as a death sentence.

Yet a forest can survive a constant number of trees. Individual trees grow and die but the forest can survive without spreading. Why can't this be true for our societies? Is growth an absolute prerequisite for survival?

The answer to this question lies in the fundamental thermodynamic difference between forest and society. This is where the notion of entropy comes in. The functioning of every living creature is accompanied by an external entropy release such as, for instance, CO_2 molecules originating from the combustion of sugars.

In the case of forests, this same release is compensated by a photosynthesis process, which generates the absorption of atmospheric CO_2 and the production of sugar that trees need in order to function. This mode of economy is sustainable thanks to solar energy and to the presence of water. Therefore, the entropy of the biosphere remains stable; it even has a tendency to decrease in geological time scales by the storage of part of the organic waste resulting from the death of the trees that make up the forest.

However, our societies are not long-lasting, as they do not produce the fuel they need in order to function. Nothing compensates the entropy rejections that result from their functioning. A mostly well-known example (but not the only one) regarding entropy rejection, is the greenhouse gas emission produced by the combustion of fossil fuels, the stocks being produced in geological time scales.

Our societies can only live at the expense of the environment, as they increase its entropy. Modern societies, mostly, are extremely harmful. Rejected entropy per capita is 100 times higher than in primitive societies, a situation that endangers their very survival. The term "Growth" used in economy is actually no more than an increasing exploitation of environment frequently called development, which is ultimately unsustainable in its present form.

There is a link between the heroes of Ancient Times and Modern mankind: in order to maintain growth, boundaries should be continuously deployed. But boundaries are not necessarily geographical. Thus a change of paradigm is necessary with the need to switch from an economy of exploitation to an economy of entropy.

CONTENTS

1

The Limits to Growth: From Malthus to COP 21 via the Club of Rome

1. Three severe warnings

Two centuries separate two events that have marked our perception of the future. The first one is the 1798 publication of Malthus' *Essay on the Principle of Population,* and the second one is the COP 21 Agreement in 2015. Between these two dates the Club of Rome's report emerged, *The Limits of Growth,* published in 1972, which fills an intermediary role, as much for its date of publication as for its content.

The link between Malthus' book, the Club of Rome's Report and the COP 21 Agreement is that they all identify a global issue of concern, and propose measures that could prevent a major socio-economic crisis in the future.

According to any of these three theories, this crisis would occur, as long as unsustainable modes of development persist. It is, indeed, exceptional that these three warnings mention the breakout of an acute crisis in the 21st century unless something serious is done to avert it.

However, these three approaches point out different reasons, as well as different methods, used to support the proposed conclusions.

2. Depletion of natural reserves or rising temperatures?

According to Malthus and the Club of Rome, the exhaustion of natural resources (arable lands for Malthus, and raw materials for the Club of Rome) will be the leading reason for collapse when meeting the needs of excessive overpopulation becomes impossible. Nevertheless, COP 21 focuses on the harmful effects of greenhouse gas emissions resulting from the massive use of fossil fuels (coal, oil and gas), their reserves being implicitly considered limitless within the given time scale. These two contrasting approaches are, indeed, striking. We are entitled to ask whether the authors of the COP 21 Agreement ever studied the widely-known Club of Rome Report.

The methods applied for these three approaches are also very different. Malthus' book is a one-man study and, thus, refers to the limited demographic data available at that time. As for the Club of Rome, it has called on a small group of MIT experts who developed the "World 3" computer program in order to follow the evolution of a number of global parameters, such as the exploitation rate of natural resources, population, industrial production and pollution, taking into account feedback loops between these parameters. It is this very interaction between these feedbacks that implies a predicted global collapse.

COP 21 conclusions rely on the work of thousands of experts who, for several years, have focused on calculating the future rise in global temperature that would result from the increase in the atmospheric concentration of greenhouse gas emissions, according to studies performed by the IPCC (Intergovernmental Panel on Climate Change). These are mainly CO_2 emissions produced by fossil fuels combustion. A rising temperature would indicate an intolerable impact if it were to exceed 2°C or even 1.5°C, with reference to the start of the Industrial Era.

2.1. *Time scales*

Although the global collapse feared by the Club of Rome and the climate crisis predicted by IPCC experts are both predicted for the 21st century, there is a sharp difference between these two studies. It concerns time estimates: how much time is left to take counter measures? Operationally, this difference is essential. If no remedial action is taken, the Club of Rome has predicted a collapse around 2015 to 2020. We will return to this point later. Implementation of the measures recommended by COP 21 is planned for 2020. Therefore, if the Club of Rome is right, then it is too late for those measures to be efficient. Who are we to believe?

Thus, a detailed comparison between these two approaches is necessary, to determine whether there is a real state of emergency or not, and if there is, on what time scale? This point will be dealt with in this chapter.

3. From Malthus to the Club of Rome

The common core of Malthus' *Essay on the Principle of Population* and *The Limits to Growth* published by the Club of Rome, is rooted in the idea that the limited resources offered to us by Mother Nature, as well as a lack of human forethought, might lead us, at some point, to a drastic crisis implying a sudden decline of population on the globe.

It is interesting to note that these two books were published during the eras when ruled by a spirit of optimism that was considered a red light by the authors.

Malthus' essay dates from the French Revolution, which bore the promise of a better society that would give everyone the chance of a prosperous life. Malthus believes that this prospect encouraged the birth rate, leading to an unsustainable increase in population that began facing the limited possibilities of increasing agricultural production.

The Club of Rome Report was published at a time when Western society was experiencing an era of unprecedented expansion, with no indication that it might stop in the near future. However, *The Limits to Growth* warns that the restriction of available natural resources might, in fact, bring this expansion to a sudden end in a rather not-so-distant future.

It is interesting to further consider the proposed arguments included in these two theses.

3.1. *Malthus' thesis*

Malthus' thesis is based on two assumptions: the geometric growth of the population, doubling every twenty-five years according to the then-available statistics; and the expansion of agricultural production, which can only be arithmetical, increasing by a constant amount at each interval of time.

3.1.1. Malthus' food crisis

Malthus concludes that the growth of population will necessarily outstrip agricultural production at a certain point in time. This situation would lead to a decrease of the food quota per capita, and, possibly, to a sharp drop in population when food quotas fall below the bare minimum. Malthus' calculations predicted this situation would occur around 2000. To avoid this scenario, he proposed a series of restrictive measures intended to limit the birth rate. In fact, according to his view, birth control needs to be imposed, because it is contrary to human nature. He believes that the development of handcraft and industry is an aggravating factor, because it reduces the number of workers available for agricultural work.

3.1.2. Malthus' thesis put to test

Malthus never justified his two basic assumptions: the geometrical growth of population, and the arithmetical growth of agricultural production.

World population was estimated at one billion at the time of publication of his essay. As shown in Table 1, it took in fact about a century until it doubled, another half a century to double again, and another half century until it doubles once again. It will reach 8 billion around 2026, according to recent forecasts. Thus, the world population has not doubled every twenty five years, but at the stage of its most rapid expansion, between the 1920s and the 2020s (forecast), it doubled every fifty years. Therefore, Malthus' assumption of a geometric growth of the population has proved itself to be about right.

However, a more detailed look reveals variances in different regions of the globe. Growth continues to double every fifty years in some regions, such as Africa, while it has slowed down (or even stopped) in others, either spontaneously — as in Europe or Japan — or due to an imposed policy, such as recommended by Malthus — in China, for instance.

Table 1.1. The world population in billions of inhabitants during the industrial era.

Population	1	2	3	4	5	6	7	8
Year	1804	1927	1959	1974	1987	1999	2012	2026

However, Malthus' main thesis, predicting that the population increases faster than agricultural production, is obviously misleading.

As shown in Fig. 1.1, from 1960 to 2010 the population increase was somewhat slower than agricultural production growth. According to a United Nations document, this trend might persist or even intensify in future decades, taking into account the slowing down of population growth.

Malthus could not have foreseen the increased agricultural production that has led to this situation of abundance. It is not due mainly to extended areas of cultivation, but rather thanks to the industrialization of agricultural processes through mechanization, and the massive use of fertilizers and pesticides. Despite Malthus' fears, the development of industry has had a positive impact on agricultural production. Nevertheless, it should be noted that the use of fertilizers has expanded faster than agricultural production: during the period from 1960 to 2010, agricultural production increased by a factor of three, while the use of fertilizers has multiplied by a factor of six, from 30 million tons to 180 million tons.

Global population and food supply — 1961 to 2051

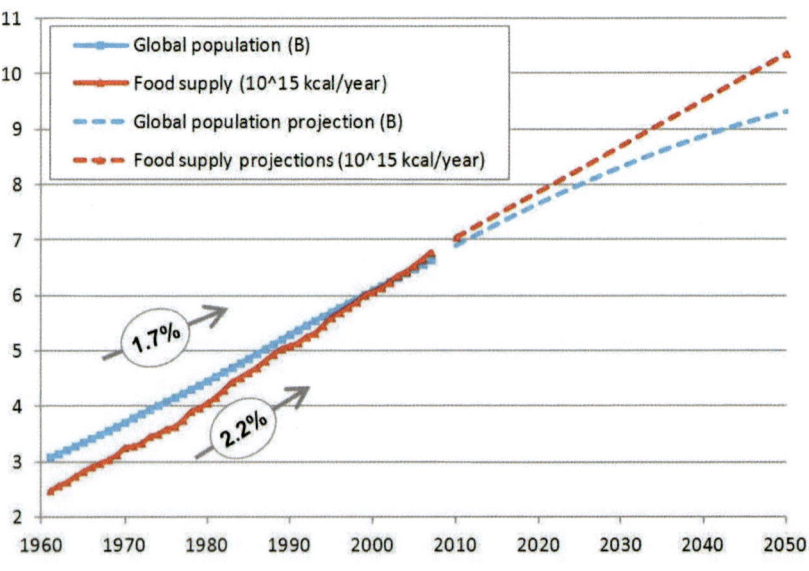

Fig. 1.1. To date, food production has increased faster than population. (*Population in billion inhabitants.*)

In general, one can say that the core assumptions of Malthus' main theses have not been proved right so far. On the contrary, there are signs of a gradual slowdown in the population growth process, and the food quota has — so far — increased. However, the massive use of fertilizers and pesticides that enabled this increase might become less and less effective, and lead eventually to a food quota decrease.

3.3. *The Club of Rome model*

There is a clear connection between Malthus' thesis and the Club of Rome Report, since, in both cases, the starting point is the finitude of natural resources. *The Limits to Growth* is obviously more ambitious. Beyond the assessment of the population growth process and the expansion of agricultural production, it takes into account other factors such as remaining natural resources, industrial production, industrial global capital, services and — an innovative factor — pollution.

The applied method is also very different. The evolution of these variables over time is calculated according to various scenarios based on a different selection of parameters such as the primary level of natural reserves, efficiency of the industrial production system, and the pollution rate for a specific production.

Pollution plays an important innovative role in comparison to Malthus' thesis. For the first time, it is mentioned that the impact of human activities on the environment might obstruct growth. However, the report does not provide any accurate definition of pollution that would enable a quantitative comparison.

3.3.1. Feedback loops and collapse

The variables are interconnected by feedback loops. The use of the World 3 program allows to study the evolution of these variables in selected scenarios containing a large number of these loops.

Some loops tend to bring the system back to balance, while others encourage instability. For instance, a negative feedback loop that brings the system into balance is that which connects pollution to population: pollution reduces life expectancy, which involves a drop in population, along with pollution for which population is responsible.

A positive feedback loop is, for instance, a loop that connects global warming and electricity consumption. In order to reduce warming effects, consumers increase their use of air conditioners, which requires an increase in electricity production, and therefore implies a rise of temperature where electricity is produced, at least partially, by fossil fuels. This loop is dominant in regions where average temperatures are high. It creates a prompt decrease of natural resources and thus leads to collapse. On the other hand, in colder regions, heating economy is the one to prevail.

3.3.2. Three limits to growth scenarios?

In 2008, Graham Turner published a comparison between the evolutions of a number of variables calculated for different scenarios based upon *The Limits of Growth* and those effectively observed during the thirty years since its publication. This comparison covers three scenarios: The *Standard Run*, or *Business As Usual* scenario, the *Comprehensive Technology* scenario and the *Stabilized World* scenario.

The *Standard Run* scenario implies a situation for which parameters reflecting physical, economic and social relations are maintained with their characteristic values or the period 1900–1970.

The *Comprehensive Technology* scenario assumes that resources are indeed extended, with 75% of materials being recycled, the level of pollution is reduced to 70% of its value in 1970, and the surface area of cultivable lands and their yield, doubled.

In the *Stabilized World* scenario, elements of social policy are added to these technological means such as: number of children per family limited to two; consumption centered on the use of services (education, health) rather than on physical goods; control of pollution; and the streaming of capital toward maintenance of land, rather than industry. These elements of social policy imply some of the ideas proposed by Malthus.

Variables such as available remaining natural resources, global population, industrial product per capita, food quota per capita, and level of pollution, are calculated for the period 1970–2000 for these three scenarios. They are then compared to those observed in the available statistics. Despite the term "natural resources" generally meaning all raw materials, Turner takes into account only fossil fuel reserves as being the most crucial raw material.

3.3.3. The limits to growth scenarios compared to the evolution observed for population, food quota and industrial production

Among the variables studied by Turner, we have selected here those that are clearly defined and therefore most appropriate for comparison with available statistics.

The first variable, and the most important according to Malthus' approach, is the evolution of the population. The evolution observed from 1970 to 2000 is compatible with either the *Standard Run* (or *Business as Usual*) scenario or the *Comprehensive Technology* scenario (Fig. 1.2). It excludes however the *Stabilized World* scenario. In the *Standard Run* scenario, the population peaks at about 8 billion in 2030 and returns in 2100 to 1970 values. In the *Comprehensive Technology* scenario, the population peaks at approximately 10 billion around 2050 (a value that corresponds to the projections shown in Fig. 1.1) but then undergoes a rapid decay before recovering, in 2100, a value comparable to that of the *Standard Run* scenario. If the Stabilized World scenario had occurred, we would have already reached a value close to the 2000 saturation, which is clearly not the case.

The second variable is the food quota. Its evolution is only compatible with the *Standard Run* scenario (Fig. 1.2). This scenario forecasts a rapid decline after 2020. The expected drop is spectacular, with the food quota already decreasing by half in 2050 to reach approximately its 1900 value. This can be considered, then, as a genuine collapse.

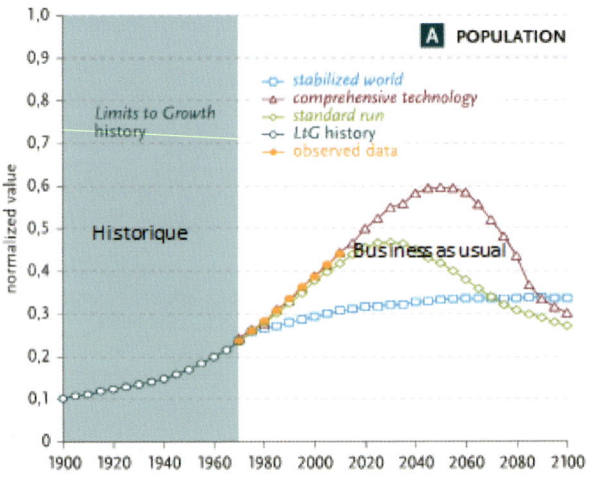

Fig. 1.2. The population is a little more than doubling between 1900 and 1970, when it reaches about 3.6 billion. The *Standard Run (Business as Usual)* scenario and the Comprehensive Technology scenario create, indeed, values measured between 1970 and 2000, but they diverge at this point. The *Standard Run (Business as Usual)* scenario predicts that the population will reach its maximum around 2030 with 8 billion, and then recover in 2100 with a value similar to that of 1980. In the *Comprehensive Technology* scenario, it reaches about 10 billion in 2050, and in 2100 recovers a value similar to that reached in 1980.

The third variable is industrial production per capita (Fig. 1.3). Only the *Standard Run* scenario appears compatible with the data statistics. For this variable, the forecasts of *Limits to Growth* are even more spectacular than for the variable "food industry quota." They provide for maximum production by 2015, followed by a rapid decline. Its value in 2030 returns finally to that set in 1970 and this after doubling from 1970 to 2015. At the end of the century, it is significantly lower than it was in 1900.

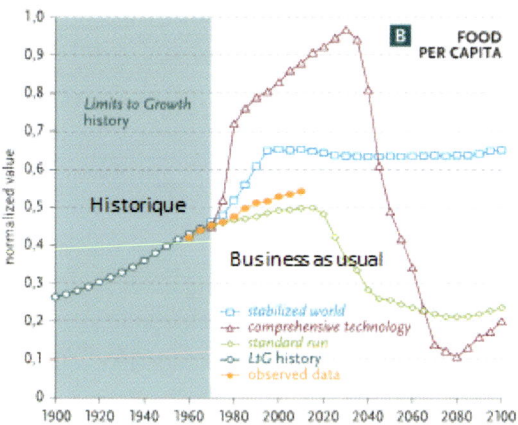

Fig. 1.3. The evolution of the food quota from 1970 to 2000 is only compatible with the *Standard Run* (*Business as Usual*) scenario. The prospects for this scenario, beyond 2020, are fundamentally different from those provided by the United Nations (Fig. 1.1). While the latter forecasts a continuous increase until 2050, the *Standard Run* (*Business as Usual*) scenario predicts a rapid decline after 2020.

3.3.4. Pollution according to Turner

Pollution as a variable plays a crucial role in Turner's clarification. While in the original version of *The Limits to Growth*, this same variable was rather inappropriate for a quantitative comparison, Turner links it to the CO_2 atmospheric concentration. Thus, he follows the trend that gradually turns this concentration into the supposed key to climate change.

In 1900, pollution variable is taken as null value; the same year, CO_2 concentration (300 ppm) was still close to that observed before the start of the Industrial Era (280 ppm). Its value in 1970 is indexed to the measured increase of CO_2 concentration at this date compared to 1900. The evolution of pollution as a variable follows quite well the one provided for by the *Standard Run* scenario. According to this scenario, pollution is still low when industrial production collapse occurs. In other words, it has a minimal role in this collapse. On the other hand, pollution becomes more crucial in the *Comprehensive Technology* scenario, although this same scenario provides for a much lower increase in CO_2 content than that observed (Fig. 1.4).

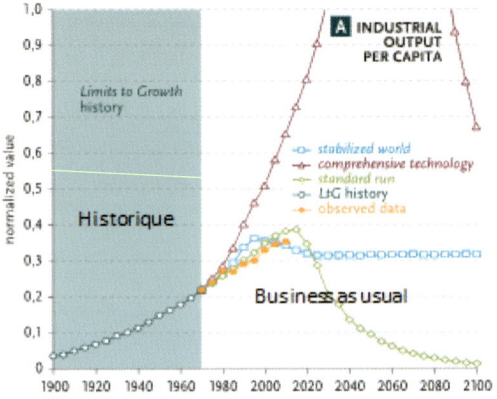

Fig. 1.4. With regard to industrial production, only the *Standard Run (Business as Usual)* scenario is compatible with the statistical data from 1970 to 2000. After doubling from 1970 to 2015, industrial production would return to its 1970 value by 2030, and in 2050 would recover its 1930 value.

3.3.5. Will the expected collapse occur?

The Limits to Growth has encountered fierce criticism due to its misleading predictions. But as Jancovici pointed out first, this report did not predict anything at all. Its sole purpose was to study the evolution of different observable variables involved in a number of possible scenarios.

   With the test of time, it turns out that only one of the scenarios described in *The Limits to Growth* — the *Standard Run* scenario — has, indeed, depicted the evolution of these variables for the period 1970 to 2000. This study, which was supposed to be no more than mere research, became a set of testable forecasts. The most spectacular ones include the collapse of industrial production by 2020, and a drastic decrease of the food quota from that date. Thus, *The Limits to Growth* model should show how valid it is very soon. For the time being, we can note a recent saturation of industrial production (Fig. 1.5). On the other hand, the food quota has continued to increase.

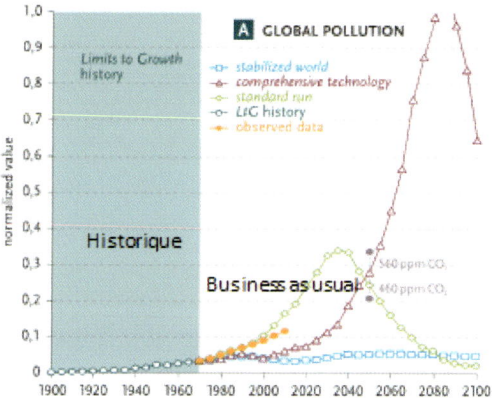

Fig. 1.5. In the *Standard Run* scenario, pollution as a variable would reach its peak in 2040, well after the collapse of industrial production expected since 2015.

However, two reservations concerning the inevitability of a close collapse should be made. In *The Limits to Growth* model, the dominant factor that triggers the collapse is the decrease of available natural reserves. According to the authors, when these stocks fall below 50% of the initial reserves, an increasing proportion of available capital must be invested in order to continue their exploitation at the same rate. In the end, it is the decreasing available capital that leads to the collapse of industrial production, as this capital is no longer available for the investment needed for growth.

When will the 50% level be achieved? The answer depends on the value of primary reserves, which is poorly known. As indicated in Fig. 1.6, the 50% level may be reached between 2020 and 2080, depending on the value of these primary reserves. The hypothesis for a collapse in 2020 is low (60,000 × 1021 joules). It assumes that only high-quality coal would be exploited, while larger reserves of low quality coal remain unused for ecological reasons.

World Industrial Production

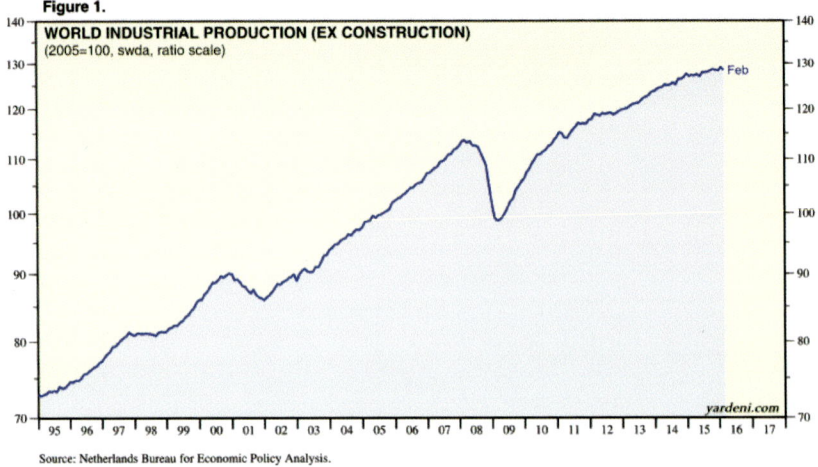

Figure 1.

WORLD INDUSTRIAL PRODUCTION (EX CONSTRUCTION)
(2005=100, swda, ratio scale)

Source: Netherlands Bureau for Economic Policy Analysis.

Fig. 1.6. After a rapid growth till 2008, industrial production shows signs of progressive saturation.

The second reservation is that the 50% level becomes irrelevant if a great part of consumed energy is renewable. Does it then prevent the risk of collapse? In fact, nothing could be less likely, since the same reasoning applies here, too: implementing renewable energies will imply the diversion of a part of available capital, just as in the case depicted by the Club of Rome, where continuing to extract fossil fuels that are less and less accessible requires more capital. Moreover, whether these renewable resources are being exploited due to the growing depletion of consumable natural resources, or because of a political decision intended to limit the use of fossil fuels or nuclear energy, does not change anything in terms of capital needed for this implementation phase. The consequences will be the same. The available capital share used just to ensure the maintenance of the production of energy will no longer be available for development.

This should be taken into account for any form of energy transition. Renewable energies that are developed at an excessive pace might lead, due to capital cost, to a collapse of the economy in the same way as the rapid depletion of natural resources in the model of *The Limits to Growth*.

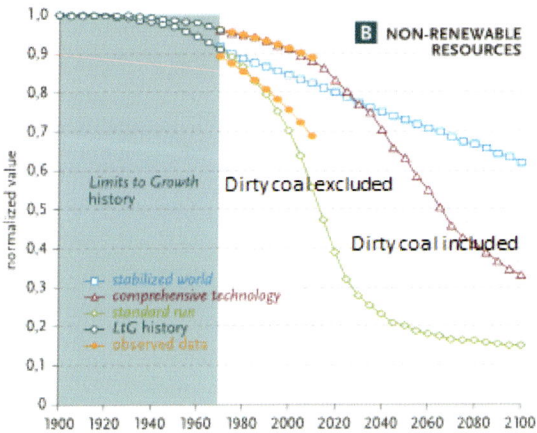

Fig. 1.7. Depletion of fossil fuel reserves depends on their initial value.

When low-quality coal reserves are excluded, the 50% level — below which the maintenance of production requires more capital — would be reached by 2020. If these reserves are included, this level would be reached fifty years later.

The predictable physical depletion of natural resources is the basis for Malthus' thesis, as well as for the Club of Rome Report. However, in his recent assessment about low usable fossil reserves, Turner points out an ecological consideration of a different nature. It reflects on the dangers linked to the use of fossil fuels — in particular, the least efficient and the most polluting fuel (low grade lignite coal).

This same awareness has led to the development of renewable energies, the capital cost of which we have referred to, along with its impact on development. Whereas in his initial vision, the Club of Rome's Report associates *The Limits to Growth* to the depletion of *energy* reserves, there is a new concern regarding a possible damage to the environment.

It presently implies an excess of entropy, rather than a lack of energy.

This brings us to COP 21.

4. COP 21

Unlike the warnings issued by Malthus and the Club of Rome, the warning launched by COP 21 is not based on the depletion of natural resources.

The COP series thus marks a break with the industrial cycle that began in the early 19th century (when Sadi Carnot wrote his essay, which we will refer to in the next chapter), a cycle mostly focused on obtaining maximum yield in order to exploit natural resources in the most optimal way.

Attention is now centered on the rise of temperatures, which, according to IPCC experts, is the result of an increase of the CO_2 concentration in the atmosphere. While the Club of Rome Report, *The Limits to Growth*, stresses the finitude of these resources and the best way to exploit them, this issue is totally absent in the commitments made by the COP 21 parties.

Here are some of the main committments taken in the Paris Agreement:

Article 2

1. This Agreement, in enhancing the implementation of the Convention, including its objective, aims to strengthen the global response to the threat of climate change, in the context of sustainable development and efforts to eradicate poverty, including by:
(a) Holding the increase in the global average temperature to well below 2°C above pre-industrial levels and to pursue efforts to limit the temperature increase to 1.5°C above pre-industrial levels, recognizing

> *that this would significantly reduce the risks and impacts of climate change;*
> *(b) Increasing the ability to adapt to the adverse impacts of climate change and foster climate resilience and low greenhouse gas emissions development, in a manner that does not threaten food production;*
> *(c) Making finance flows consistent with a pathway towards low greenhouse gas emissions and climate resilient development.*
> *2. This Agreement will be implemented to reflect equity and the principle of common but differentiated responsibilities and respective capabilities, in the light of different national circumstances.*

In this article, the parties undertake to limit the increase in the heating of the planet to 2°C, or even 1.5°C (2.1a), by promoting low greenhouse gas emissions (2.1b). The parties admit that the achievement of this goal will require a change in the process of capital flow (2.1c).

> *Article 4*
>
> *In order to achieve the long-term temperature goal set out in Article 2, Parties aim to reach global peaking of greenhouse gas emissions as soon as possible, recognizing that peaking will take longer for developing country Parties, and to undertake rapid reductions thereafter in accordance with best available science, so as to achieve a balance between anthropogenic emissions by sources and removals by sinks of greenhouse gases in the second half of this century, on the basis of equity, and in the context of sustainable development and efforts to eradicate poverty.*

In this article, the parties undertake, during the second half of the century, to reach a balance between anthropogenic emissions and the absorption of greenhouse gases within a sustainable development process. In other words, it means that, from 2050, net emissions of CO_2 shall be zero.

Article 6

8. Parties recognize the importance of integrated, holistic and balanced non-market approaches being available to Parties to assist in the implementation of their nationally determined contributions, in the context of sustainable development and poverty eradication, in a coordinated and effective manner, including through, inter alia, mitigation, adaptation, finance, technology transfer and capacity-building, as appropriate. These approaches shall aim to:
(a) Promote mitigation and adaptation ambition;
(b) Enhance public and private sector participation in the implementation of nationally determined contributions;

In this article, the parties recognize that the set goals cannot be achieved if based on market mechanisms only. The parties therefore take note of the fact that the implementation of renewable energies will require investment that would be considered unprofitable according to the usual criteria of liberal economy.

Article 12

Parties shall cooperate in taking measures, as appropriate, to enhance climate change education, training, public awareness, public participation and public access to information, recognizing the importance of these steps with respect to enhancing actions under this Agreement.

This article is consistent with Article 6, since the financing of the transfer to an economy implying energy production without CO_2 net emissions will require the active intervention of public authorities. It will be necessary to convince the taxpayer (who will bear the cost) that this intervention is worth it.

Article 13

7. Each Party shall regularly provide the following information: FCCC/ CP/2015/L.9 29 (a) A national inventory report of anthropogenic emissions by sources and removals by sinks of greenhouse gases, prepared using good practice methodologies accepted by the Intergovernmental Panel on Climate Change and agreed upon by the Conference of the Parties serving as the meeting of the Parties to the Paris Agreement; (b) Information necessary to track progress made in implementing and achieving its nationally determined contribution under Article 4.

This article takes note of the fact that the commitments made by the parties concerning CO_2 emissions cannot be easily independently verified. The reason for this is that CO_2 emissions are rapidly homogenized on a global scale in accordance with a fundamental physics phenomenon (Boltzmann's law), which will be discussed in the next chapter.

The COP 21 Agreements form a coherent whole, provided we accept that a rise of temperatures above 2°C, or even 1.5°C, represents a major risk for humanity. These agreements are based on the scientific research provided by the IPCC.

The parties recognize that the implementation of these agreements will require significant investment that will not be covered by market mechanisms. It leads to two consequences. First, relevant capital will not be available for other investments; this is a new type of limitation to growth that has nothing to do with the finitude of natural resources. On the other hand, technology thus becomes secondary. In fact, the foundations of liberal economy will be challenged.

5. Should these three warnings be taken seriously?

The dangers put forward by Malthus have not materialized so far. Food quota has been steadily improving over the last fifty years, and is expected to carry on this way, according to the UN forecast, at least until the middle of the 21st century. There is no imminent concrete threat of large scale famine. The spontaneous evolution of demography in developed countries also shows that an exponential growth of the population is not a fatality, contrary to Malthus' theory.

The starting point of *The Limits to Growth* — collapse of the world economy resulting from the depletion of natural resources, mainly fossil fuels, according to Turner — seems to have been overturned as well. Human ingenuity has considerably extended the field of exploitable resources. There is still a lot of oil, gas and coal.

Nevertheless, there is a striking similarity between the evolution of industrial production observed between 1970 and 2000 and the one predicted by the *Business as Usual* scenario. The reason for this similarity remains unclear. However, if the *Business as Usual* scenario were to apply, we would be on the eve of the predicted collapse. Is the absence of growth observed on a world scale over the past few years, the prelude? Or is it the sign of progressive adaptation?

But what does this adaptation mean? The COP 21 Agreements reflect a wave of public opinion calling for the end to exploitation of the fossil fuels that generate CO_2 emissions. They are almost unanimously considered to be the cause of observed global warming. This public trend reveals a sharp change. It is the very notion of natural resources that is being transformed. From an objectively measurable quantity, natural resources become a value quantified by a political choice, as already mentioned by Turner when he excludes low quality coal from these polluting reserves in the *Business as Usual* scenario. It is this choice that triggers the 2020 collapse in this scenario.

But is global warming as big a threat to humanity as is claimed by the COP 21? It is a relevant question as the end of the exploitation of fossil fuels, either partial or total, might also represent a danger, since their replacement by renewable energies might devour a major part of available capital. This would trigger a fall in growth and industrial production — and consequently a drop in food quota — as Turner has shown.

Therefore, a thorough understanding of the reasons that prompted COP 21 to recommend massive reduction of greenhouse gas emissions is essential.

This is the topic of the next chapter.

2

Greenhouse Gas Effect and Boltzmann Entropy

In order to function, any living organism requires an energy supply which is fuel. If fuel runs out, life stops. This is the very core of Malthus' theory, as well as the central element in the model used by the authors of *The Limits to Growth*. Such works recognize that our societies' characteristics rely on the tendency to grow, implying an increasing use of natural resources. Over time, the exhaustion of these resources leads to the collapse of societies.

But the very functioning of any organism also produces emissions that pollute the environment. Heavy pollution might be life threatening, even if resources are not exhausted. Therefore, these polluting elements impose a growth limit, of a different kind from that due to the finitude of natural resources.

These elements might involve air, land or water pollution. In terms of thermodynamics, all types of pollution lead to an increase of environmental entropy. Any living organism grows and exists while extracting energy from its environment and rejecting entropy. It exploits and degrades energy at the same time.

As presented in the previous chapter, the term "pollution" is explicitly included in *The Limits to Growth*. It is a new element and an essential step with regards to Malthus' theory which focuses only on the finitude of natural resources. However, the Club of Rome Report does not link the specific case study (soil pollution reducing agricultural yields, thus limiting the population that can be fed) with the much more global issue concerning emissions that increase environmental entropy.

There is also a striking contrast between Malthus' theory and the Club of Rome Report on the one hand, and the COP 21 Agreements, on the other. The Agreements do not even once mention the fact that finitude of resources might limit growth. It focuses on greenhouse gas — mainly CO_2 — as being the imminent threat. Implicitly, and perhaps unknowingly, COP 21 points out the threat of entropy, instead of the energy threat raised by the Club of Rome Report, which considers fossil fuels to be the essential natural resource.

1. CO_2 — considered as waste

The main issue focuses on CO_2 as harmful waste. In this regard, COP 21, like previous COPs, adheres to Turner, who prefers to link the "pollution" variable to the atmospheric concentration of CO_2. However, for Turner, this variable plays a relatively minor role. It is not the increase in pollution that leads to the collapse of the system according to the most probable *Standard Run* scenario in *The Limits of Growth*, as can be seen in Fig. 1.5. In this scenario, the level of pollution collapses after 2040, well after the industrial production era. *The Limits to Growth* underlines the decline of industrial production that leads to the decrease of pollution (precisely, CO_2 emissions), and not the other way around.

On the other hand, COP 21 considers natural reserves to be limitless. It is the increase of CO_2 that constitutes the main threat. We have gone from an era in which depletion of natural resources — mainly fossil fuels — was considered the leading problem, to the present era in which the accumulation of CO_2 waste creates the most severe threat.

The concentration of CO_2 in the atmosphere has become the landmark and the measurement to rate pollution, although, in reality, it is only part of it. How did we get there?

1.1. *CO_2 emissions and evolution of the temperatures*

The Paris Agreements are the outcome of a process that began in the 1980s with the first signs of possible global warming. Indeed, what started as a mere hypothesis has become a widely-recognized fact. The average temperature of the globe increased by less than 1°C since the start of the Industrial Era, but the main focus of this increase seems to be on the last thirty years, as can be seen in Fig. 2.1.

This figure indicates the difficulty to pinpoint the start of anthropogenic warming because of temperature fluctuations. The warming process since 1980 looks very clear. However, a further look shows a slight cooling from 1880 to 1910, followed by a warming up from 1910 to 1940, then a plateau from 1940 to 1980. Detailed simulations are required in order to distinguish anthropogenic warming from the natural variations that are influenced by a series of factors, such as solar activity.

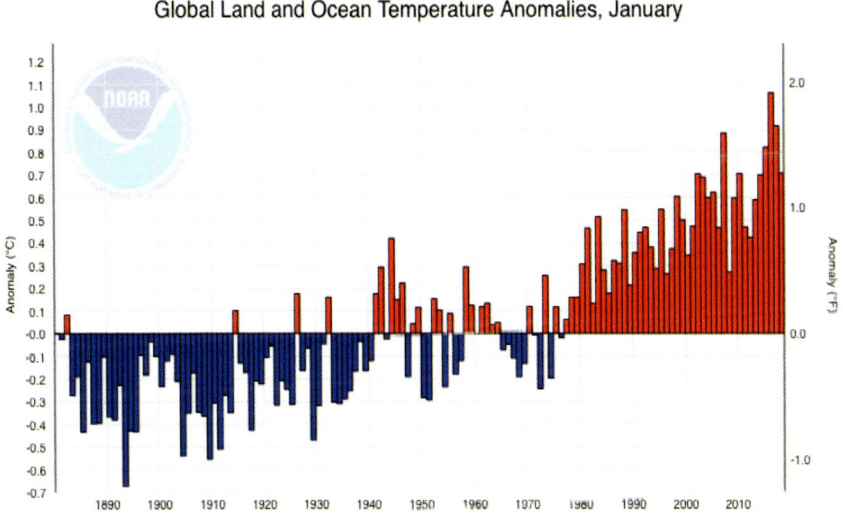

Fig. 2.1. Temperature anomaly records since 1880.

The anthropogenic increase is generally linked to the atmospheric concentration of greenhouse gases, mainly CO_2, which has been dramatic since the 1950s, as can be seen in Fig. 2.2.

However, we note a gap of about thirty years between the start of this rapid increase and the rise of temperatures: emissions have increased rapidly since 1950, while temperatures did not rise until about 1980.

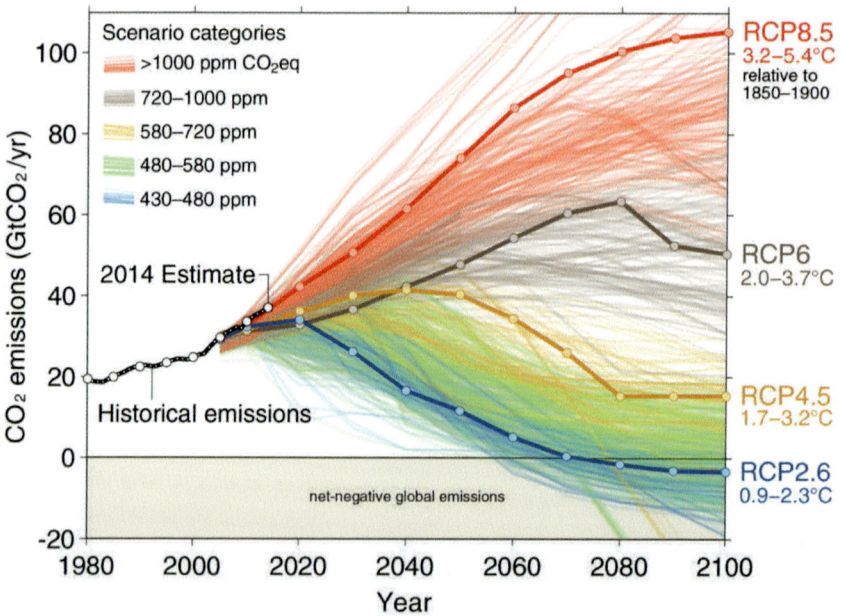

Fig. 2.2. Evolution of carbon emissions for different scenarios ranging from *Business as Usual* (RCP 8.5) to *Stabilized World* (RCP 2.6). The white dots represent average emissions up to 2010. Beyond this, the curves represent four different emission scenarios. In the box, the corresponding CO_2 concentrations can be seen.

As previously done by the authors of *The Limits to Growth*, IPCC experts have studied different CO_2 emission scenarios that correspond to various levels of the use of fossil fuels. In 2100, according to the RCP 8.5 scenario (*Business as Usual*), the CO_2 concentration will have reached 900 ppm and emissions amount to 25 PgC (Petragramme carbone)/year. In the RCP 2.6 scenario

(Stabilized World), emissions fall to zero and CO_2 concentration falls back to its current level, slightly higher than 400 ppm.

Expected estimates of temperature increases for 2100 are presented in different scenarios studied (Fig. 2.3). In the *Business as Usual* scenario, the average temperature would have increased by 4.5°C since 1950; in the *Stabilized World* scenario, the increase would be only 1.5°C. In this scenario, which is the one preferred

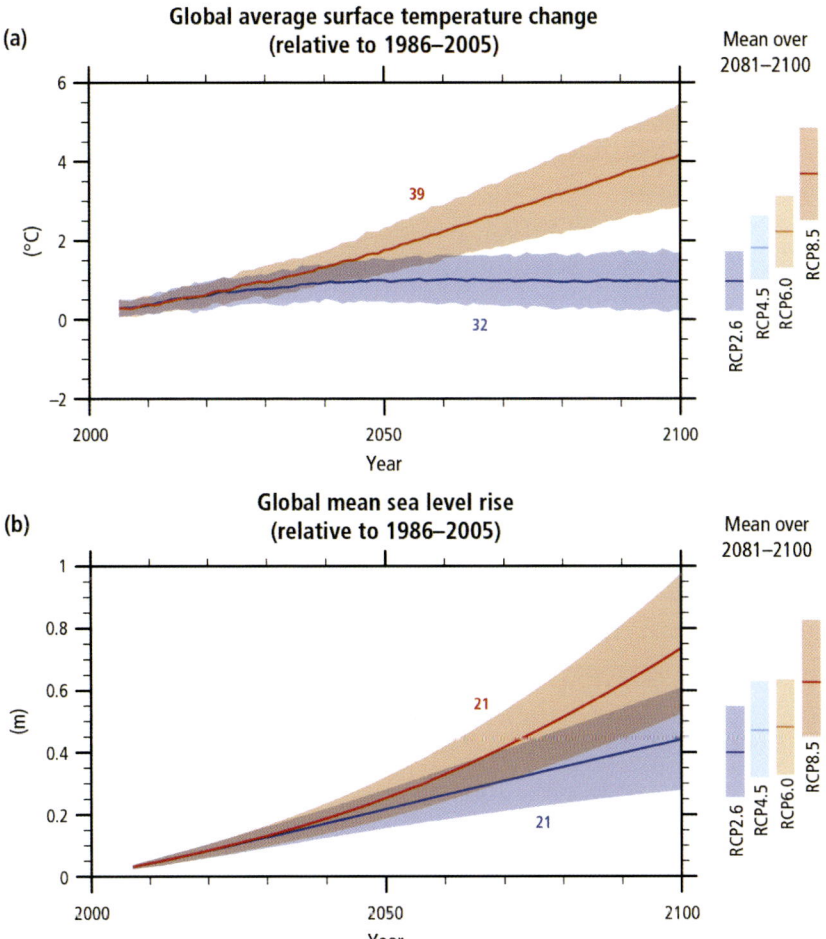

Fig. 2.3. Changes in temperatures since 1950 in scenarios RCP 8.5 (*Business as Usual*) and RCP 2.6 (*Stabilized World*).

by COP 21, emissions must be compensated by carbon sinks (also anthropogenic) from 2050 onwards. In other words, net CO_2 emissions should fall to zero at this precise date.

1.2. *Carbon reserves and IPCC projections*

IPCC projections within the various considered scenarios do not take into account depletion of carbon reserves (oil, gas and coal). Can these reserves, which are at the core of *The Limits to Growth*, slow down the predicted increasing temperatures, and if that is the case, for which scenarios exactly?

Turner argues that exploitable carbon reserves, without including low-quality coal, amount to 60.10^{21} joules. Taking for combustion heat the value of 32.8 MJ/kg, we get in the units used in Fig. 2.2 the value of 2000 PgC. On the basis of this value, cumulative emissions from 1850 to 2100 can be calculated for the various given scenarios.

For the *Stabilized World* scenario (RCP 2.6) with net emissions (i.e. emissions minus removal per carbon sink) falling to zero by 2100, a cumulative emission of 675 PgC is obtained, or approximately one third of the reserves. Thus, they have no impact on this scenario. It is however hardly probable as it would require a sharp fall of emissions in the very near future. It should also be taken into account that this scenario presents a decrease of emissions resulting from the use of carbon sinks based on CO_2 stored under pressure. Thus, it will be impossible to use this carbon for fuel later on. The finitude of reserves will eventually have an impact, but only after 2100.

For the least optimistic scenario — RCP 8.5 — emissions that will accumulate by 2100 are 1 875 PgC, which is similar to the reserves value. This scenario should therefore be excluded. At this rate of exploitation, emissions will fall before 2100 simply due to a lack of fuel.

For the intermediate scenarios RCP 6.0 and RCP 4.5, emissions that will accumulate by 2100 represent about half of the reserves.

So, until then, their impact on the given scenarios will be limited. But their impact will be more concrete after 2100. With more than half of the reserves now consumed, a further rise of temperatures in the 22nd century may not take place.

Considering the RCP 6.0 and RCP 4.5 scenarios as the most probable, IPCC simulations may forecast a temperature increase of about 3°C by 2100.

In any case, finitude of carbon reserves will have an impact on temperature evolution, but this will mainly occur in the 22nd century. It is very likely that by the end of the 21st century, the temperature will exceed the values set by the Paris Agreement in Article 2a. They will continue to increase in the 22nd century.

2. Greenhouse gases, spatial homogeneity and seasonal variations

Calculations regarding the atmospheric concentration of CO_2 in different parts of the world show only slight variations according to latitude, from the South Pole to Alaska (Fig. 2.4). The difference between concentrations recorded in both hemispheres amounts to only about 2–3 ppm.

Values in the southern hemisphere follow those of the north with a delay of about one to two years. This is the lapse of time required for CO_2 molecules to move from the north, where they are mostly released, to the south.

However, the amplitude of seasonal variations is closely linked to latitude. They are very pronounced in the northern hemisphere, where their amplitude reaches 20 ppm in Alaska (Fig. 2.5), while they barely reach 1 ppm at the South Pole (Fig. 2.5d).

Moreover, in the northern hemisphere, a minimum is noted in October (Fig. 2.5b), while in the southern hemisphere, it occurs in March (Fig. 2.5c). At the Mauna Loa Observatory in Hawaii, variations show amplitude of a little less than 10 ppm, with a minimum value in October.

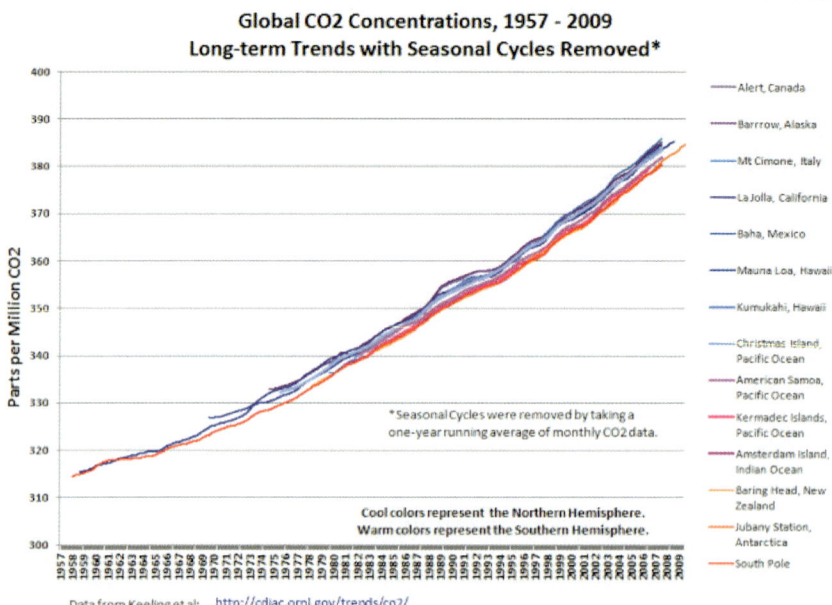

Fig. 2.4. Global CO_2 concentrations 1957–2009. Long-term trends after subtraction of seasonal cycles (in blue, northern hemisphere, in red, southern hemisphere). CO_2 concentration varies only slightly with latitude.

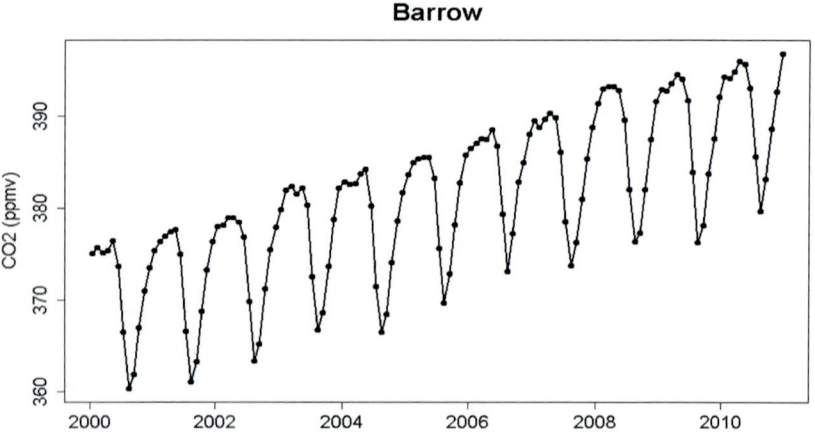

Fig. 2.5a. Measurements of CO_2 concentration carried out at the Barrow station in Alaska show an increase of about 30 ppm in 14 years, as well as a significant oscillation amplitude of around 10 ppm.

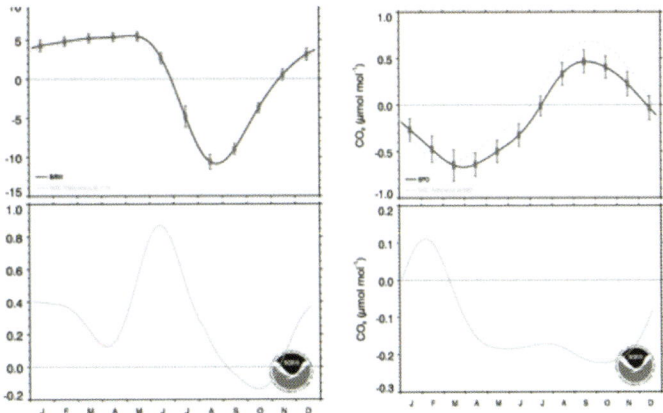

(Left) Fig. 2.5b. Great seasonal difference in Alaska, minimal in autumn.
(Right) Fig. 2.5c. Slight seasonal difference at the South Pole, minimal in spring.

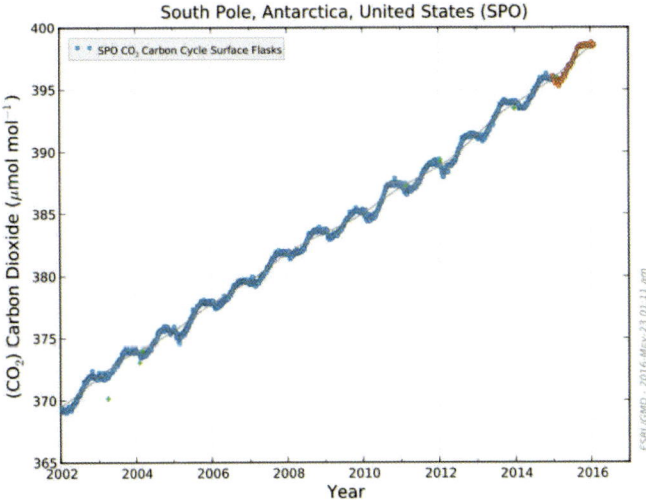

Fig. 2.5d. Measurements at the South Pole show the same increase as in Alaska,
but with much smaller seasonal variations.

These two observations — large-scale homogenization and differences in seasonal variations between the northern and southern hemispheres — are the hallmark of two fundamental phenomena.

Homogenization is a worldwide expression of Boltzmann's Law, stating that diffusion always occurs from regions with high concentrations towards regions with low concentrations. We deal at length with this law later.

Seasonal variations are marked by a decreasing concentration in summer, from April to September in the northern hemisphere, and from October to March in the southern hemisphere, with a winter increase in the northern hemisphere and a summer increase in the southern hemisphere. In summer, plants (by means of photosynthesis) transform CO_2 into sugar for growth. CO_2 concentration decreases, and oxygen concentration increases.

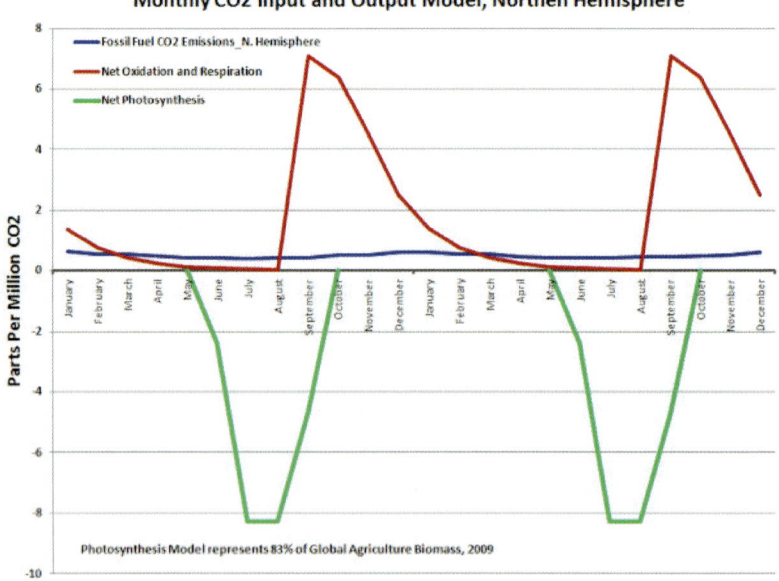

Fig. 2.6. Photosynthesis in the northern hemisphere reduces CO_2 concentration from March to October, while respiration restores it to the atmosphere from August to April.

In winter, respiration restores CO_2 to the atmosphere and draws out the oxygen. The difference of the seasonal variation amplitude between the hemispheres is due to the fact that most emerging lands — and thus, forests where most of the photosynthesis takes place — are located in the northern hemisphere. At the South Pole, where there is no vegetation, there is neither photosynthesis in summer nor respiration in winter.

Figure 2.6 shows in a separate schema, the effect in the northern hemisphere of CO_2 absorption (photosynthesis) in summer and CO_2 release (respiration) in winter. There is, indeed, a balance between these two phenomena. Regarding the time scale for which detailed measurements are available, i.e. about a century ago, CO_2 storage values by forests are not directly accessible according to this type of measure.

3. Atmospheric CO_2 measurements: the scientific basis of the Paris Agreements

The homogenization of CO_2 concentrations is, in itself, a fascinating phenomenon. It is a fundamental physics experiment on the scale of the earth, and its social consequences, which are known today, are enormous.

Regions with the lowest emissions — countries where people are the poorest — are affected in the same way as regions where these emissions are high and where people are rich. In the United States, emissions per capita are ten times higher than in India or in the Philippines, and twice as high as in Europe. China is rapidly approaching the level of European emissions. The most striking example of this "injustice" is the African continent. With 16% of the world's population, it contributes only 1% to these emissions, but CO_2 concentration and its effects are the same as elsewhere.

This homogenization, which illustrates Boltzmann's Law as we will see below, is one of the scientific bases of the Paris Agreements.

It is the core of Article 2.2, which recognizes the *principle of common but differentiated responsibilities and respective capabilities, in light of different national circumstances.*

It is also the reason for Article 13, whereby the parties undertake to provide on a regular basis: ... *a national inventory report of anthropogenic emissions by sources and removals by sinks of greenhouse gases.*

These relationships are necessary since homogenization is unable to set the contributions originating from different countries via independent measures.

Figures 2.2 and 2.3 show that only the *Stabilized World* scenario can limit warming to 1.5°C. on a permanent basis, as advocated by the Parties (Article 2a). This scenario requires the total eradication of emissions after 2050, as provided for in Article 4: *In order to achieve the long-term temperature goal set out in Article 2, Parties aim to achieve a balance between anthropogenic emissions by sources and removals by sinks of greenhouse gases in the second half of this century...*

All the articles of the Paris Agreements are therefore coherent and rest on a sound scientific basis.

4.   Required investments for realization of the carbon-free stabilized scenario

Articles 2c and 6 take note that a total eradication of emissions after 2050 in the *Stabilized World* scenario will require investments that cannot be attained by market mechanisms. It is therefore recognized that they will bear no profit in the traditional economic sense. They will mainly be supported by the state — in other words, by the taxpayer.

The Paris Agreements avoided any estimation of these investments, as it is obviously a thorny issue and involves uncertainties that are barely mentioned in these agreements. However, this subject must be pointed out if we want to remain realistic. The IPCC thoroughly examined the adverse consequences of global warming, but the Paris Agreements based on these studies do not indicate what financial means will be required to avoid this warming.

One can, however, easily get an idea of the investment to be made before 2050 within the Stabilized World scenario. Investment in renewable energies for the year 2015 amounted to US\$ 286.10^9, for an installed power capacity of 150 GW. The installation of one renewable GW costs therefore about US\$ 2 billion.

The installed electrical power in the world amounts to approximately 5,500 GW; investment required to make electricity production carbon-free by the installation of renewable energies reaches thus US\$ 11.10^{12}. However, it should be borne in mind that renewable energies provide electricity for only 30% of the time. To produce the same energy as conventional power plants, that produce electricity all the time, an investment of US\$ 33.10^{12} would be necessary. In fact, this would still not suffice, because part of the energy produced should be stored to ensure permanent distribution, or one would have to keep in stock power plants operating with fossil fuels or nuclear energy. It would also be necessary to ensure the transport of "green" electricity over long distances; and yet, we are dealing only with a carbon-free electricity production, which represents no more than 20% of total energy consumed.

Let us therefore suppose a sum of US\$ 100.10^{12} as the cost of having a totally carbon-free economy. This is roughly the annual global GDP to be invested over 30 years. So, every year, 3% of the world's GDP should be devoted to carbon-free energy consumption. Total carbon-free processed energy would therefore bring the net economy growth to zero.

The same conclusion applies when adopting an economy supplied by nuclear power plants rather than by renewable energies. Indeed, the cost of nuclear electricity production is now the same as that of renewable energies, meaning about €0.10 to 0.20 per kWh.

The *Stabilized World* scenario, to which the Paris Agreements Parties have committed, would therefore bear a heavy social cost. This is, in fact, the price to be paid for the limitation of entropy increase in the biosphere, while maintaining the current level of energy consumption.

Chapter 6 presents more detailed estimates of the cost of a transition involving the elimination of fossil fuels, with minimal storage of electrical energy supplied by renewable sources. These estimates also take into account GDP differences between rich and poor countries.

Global warming forces us to recognize that anthropogenic emissions limit our growth via measures that are necessary to limit global warming. These emissions increase entropy in the biosphere. This leads us to the study of the definition of entropy given by Boltzmann.

5.   Pollution and Boltzmann's Law

All forms of pollution result from Boltzmann's Law, which is fundamental to thermodynamics. According to this law, impurities introduced into the environment tend to proceed to maximal dilution. This law applies to greenhouse gas emissions, which are released into the atmosphere, and are responsible for global warming as well as fine particles that pollute the air we breathe in cities, and chemicals such as pesticides that infiltrate groundwater. As explained in the following section, this dilution increases environmental entropy, while reducing free energy. Dilution is an energetically favorable state and this is the reason why greenhouse gases tend to spread throughout the atmosphere.

We have summarized the above data showing that greenhouse gases such as CO_2 do not remain confined in the space where they were emitted. On the contrary, they spread through the atmosphere. As we have seen, measurements of their concentration in different parts of the globe show remarkable homogeneity. CO_2

molecules mostly emitted in the northern hemisphere are found one or two years later in the southern hemisphere at the same level of concentration. CO_2 pollution emissions equally affect industrial zones of the world, such as China, and regions where there is no use at all of fossil fuels, such as Antarctica. For example, Germany emits more CO_2 than France, but the effects of its emissions are the same in both countries.

This greenhouse gas phenomenon is the expression of a more general law. As proof: if we introduce a drop of dye in a bucket of water, even without shaking the liquid, the color of the water becomes homogeneous after a certain time. It is then impossible to pinpoint where the drop of dye was introduced, even if we did not shake the liquid in order to accelerate the homogenization process. Similarly, atmospheric currents are not the fundamental reason for the homogeneity of greenhouse gas concentration.

All forms of pollution are subject to the same laws. The fine particles contained in the exhaust gases of cars do not remain at ground level; pictures of Beijing on days of heavy pollution provide simple proof. In addition, chemicals used for agricultural and industrial purposes pollute the surroundings, including rivers and soils.

The reason for this homogenization trend relies on Boltzmann's Law, which we briefly explain below.

5.1. *Boltzmann's Law*

Today, the existence of molecules seems obvious, but in fact, it has been proved only recently.

The second half of the 19th century was shaken by critical scientific controversies. One of them precisely concerned the existence of atoms and molecules, which had not yet been established experimentally, although the idea had already been suggested by Democritus in ancient times. It is even counter intuitive as, on our discernible scale, matter seems to be continuous, whereas, according to Democritus, it is composed of indiscernible and indivisible elements.

The existence of molecules is the fundamental assumption of Ludwig Boltzmann (1844–1906). Based on this very assumption, his likely theory of the evolution of systems comprising a large number of identical, indiscernible and indivisible particles was harshly criticized by leading scientists such as Mack and Ostwald. Some others, like Clausius and Maxwell, supported him. But the fact that gas was composed of distinct molecules was only firmly established after his suicide. Their existence was proven through a process of observation of the erratic movements of particles suspended in fluid — the Brownian motion — combined with the theoretical interpretation of these movements by Einstein. This interpretation is based on the existence of molecules (not directly visible) that compose the fluid. Suspended particles, which are bigger than molecules, are subject to random shocks due to their thermal agitation. This enhances their erratic movement.

Boltzmann's theory predicts that any system consisting of indivisible entities spontaneously evolves toward the most probable state, which is the one containing the largest number of indiscernible microscopic states on a macroscopic scale. Thus, if CO_2 molecules in the atmosphere remained confined to their site of emission, each molecule would occupy only a small number of positions. The number of microscopic states that characterizes this concentrated state would remain low. On the other hand, if CO_2 molecules are free to occupy any position in the atmosphere, the number of microscopic states that characterizes this diluted state is much greater. According to Boltzmann, an initial state of concentration will evolve toward the most diluted state possible, as it is ultimately the most likely one. On the other hand, a diluted initial state will not evolve to a concentrated state because it is less likely. This theory is probabilistic. An improbable state might exist — but it still remains improbable.

The homogenization of the concentration of greenhouse gases in the atmosphere, wherever they have been emitted, is a wonderful, large-scale illustration of Boltzmann's Theory.

5.2. *The entropy of Boltzmann and the entropy of Clausius*

For Boltzmann, W represents the number of indistinctive microscopic configurations at macroscopic scale in a given state. Within an isolated system, Boltzmann proposes a statistical definition of entropy (S) previously introduced by Clausius:

$$S = k_B \ln(W)$$

here k_B is a universal constant with the dimension of energy split by temperature.

For a system based on a fluid with a concentration c (per unit of fluid volume) of suspended foreign molecules or particles, W is the number of positions that each one of them may occupy. The entropy per unit of volume associated with the presence of these foreign particles or molecules is then:

$$S = c\, k_B \ln(W)$$

As Clausius has shown, the energy that governs the evolution of a system is not the usual energy, but an energy called "free energy," denoted by the letter F. It is the real usable energy. It decreases when entropy increases:

$$F = U - TS$$

where U is the usual energy, T is the absolute temperature, and S is Boltzmann's entropy.

Entropy in a diluted state is greater than entropy in the concentrated state because foreign molecules or particles may occupy a greater number of positions. The free energy value within the diluted state will, therefore, be lower than in the concentrated state.

Any system left alone will result in greater dilution, because this is how it minimizes its free energy. It is the reason why the concentration of CO_2 molecules in the atmosphere tends to homogenize. The measurements shown in Fig. 2.4 are a global illustration of Boltzmann's Law.

It is interesting to recall the reasons which led Clausius to choose the term "entropy":

> "I propose to name the quantity S the entropy of the system, after the Greek word [τροπ trope], the transformation. I have deliberately chosen the word entropy to be as similar as possible to the word energy: the two quantities to be named by these words are so closely related in physical significance that a certain similarity in their names appears to be appropriate."

This magnificent sentence should be highlighted in any document dealing with the so-called energy crisis. According to Boltzmann, the International Energy Agency should, in fact, be called the International Agency for Energy and Entropy. Relevant debates would be much clearer if it was stressed from the start that these two notions are inseparable.

6. The thermal machine model of Sadi Carnot

The notion of entropy, if not the term itself, goes back to Lazare Carnot and especially to his son, Sadi Carnot — and this is perhaps the reason that Clausius chose the letter S, as he was a great admirer of Sadi Carnot.

Sadi Carnot aimed to set up a true theory regarding thermal machines that were used during the steam era. At the start of the 19th century, the beginning of the Industrial Era, they gained wide expansion in England. Their motor strength opened the way to the large-scale exploitation of underground coal and iron mines as they enabled water to be pumped from the galleries leading to these mines. The rest would follow.

In a steam engine, coal is burned to heat water that is turned into steam at high pressure. This steam activates a piston, with a movement that provides power, before being condensed again into water.

As all physicists are trying to do, Sadi Carnot reduced the problem of the thermal machine to its simplest form. He put aside details of the boiler, the condenser, the fluid which is used to convey heat from one to the other, the cylinder and the piston — all the mechanic details. His model machine comprised only three elements: a fixed (T1) temperature heat source, a fixed (T2) temperature cold source, and a fluid which carries a quantity (Q) of heat from one to the other via a device that provides (W) work.

When schematized in this way it is obvious that, in principle, this machine could also work in reverse to bring up (Q) heat from the cold source to the hot source using (W) work provided from outside. This is what we call a heat pump.

This schema immediately enabled Sadi Carnot to set up a fundamental question: for a given (W) work, can a machine operating in heat pump mode bring up to the hot source more heat than it would transfer to the cold source in thermal machine mode? For Carnot, the obvious answer was no. He had just formulated the second principle of thermodynamics (in fact the first principle, the equivalence of heat and energy would only be stated later). According to this principle, heat cannot rise spontaneously from a cold source to a hot source. It would then mean a perpetual movement. Likewise, according to Boltzmann, CO_2 molecules that are scattered in the atmosphere cannot concentrate spontaneously in one location. Or, more precisely, the probability for such phenomena to occur is extremely low.

The spontaneous passage of heat from "hot" to "cold" and the spontaneous tendency to dilute are two aspects of the same principle. In a hot liquid, the molecule's kinetic energy is higher than in a cold fluid. When both fluids mix, energies are equal and the molecules become indiscernible. After mixing, the number of indiscernible microscopic configurations increases, thus, Boltzmann's entropy increases.

One might therefore date back to Sadi Carnot the emergence of the notion of entropy, eventhough the word itself was to be invented much later by Clausius. When Sadi Carnot argues that heat cannot pass spontaneously from a cold source to a hot source, he already implies that the entropy of a closed system cannot decrease. Indeed, the greater the difference between cold source and hot source increases, the more the number of distinct microscopic configurations, W, decreases.

7.   Entropy and the Paris Agreements

In a closed system, Boltzmann's entropy can only increase. The entropy of a system can only decrease if provided with energy from the outside world. This is what happens in the Carnot machine when it operates in the heat pump mode, when the outside world provides energy to raise heat from the cold source to the hot source.

The word "entropy" does not appear in the Paris Agreements. The experts and the drafters of these Agreements do not seem to have understood what it is all about. The purpose of these Agreements is precisely to limit the increase of entropy in the biosphere, or even reduce it and bring it back to its balance, by reducing the CO_2 atmospheric to its historical value; i.e. 280 ppm. Carnot, Clausius and Boltzmann have, indeed, taught us that this would require energy, since a reduction in CO_2 concentration leads to a decrease in entropy. And this necessarily has a cost, which we have grossly estimated above. But the cost of energy is more important than the financial cost. For this purpose, we have to learn how to quantify the notions of entropy and free energy. This will be the focus of Chapter 4. But before that, we must first draw up a general picture of the evolution of the biosphere on geological time scales.

3

Biosphere in Disruption

Nowadays, climate disruption represents a major concern. Some argue that this change is due to human activities, which implies that "natural" climate would be perfectly regulated and suffer no disruption. This is not the case, however. Temperature, concentrations of CO_2 and oxygen, as well as other parameters, have undergone considerable variations long before Man appeared on Earth. Variations are minimal, today, compared to those that occurred in the distant past.

Regarding the time scale of the age of the planet, climate is the result of favorable physical conditions — mostly involving the presence of water (H_2O) in liquid form — that enabled the development of life on Earth. Without life on Earth, the concentration of CO_2 would have remained at a level of several thousand ppm, and temperatures would be about ten degrees higher than they are today. Forests, through photosynthesis, have reduced the concentration of CO_2 down to values of a few hundred — 400 ppm today — and have kept temperatures at levels compatible with the broad evolution of the animal kingdom. But would a temperature increase of 2°C truly be a threat as argued?

The COP 21 Agreements are based on the current consensus that the increase of CO_2 content from the pre-industrial era value 280 ppm, due to the massive use of fossil fuels, is precisely the cause of global warming. The argument that CO_2 acts like the glass roof of a greenhouse is well known. This glass roof allows the visible part of the spectrum emitted by the sun to penetrate the greenhouse, but it also intercepts infrared radiation emitted by soil and plants, which are redirected inside the greenhouse. As a result, the heat is trapped in the greenhouse. CO_2 molecules act just like a glass roof, because they absorb infrared radiation.

But this theory is too simplistic to be applied to the biosphere. There are many other factors to be taken into account, such as interactions between the atmosphere, the oceans and the Earth, the effects of convection in the atmosphere and a number of other elements. Thus, the current consensus relies on complex biosphere models that enable the connection between concentrations of CO_2 and temperature, while taking into account a large number of factors.

However, these models are so complex that they become inaccessible to non-experts, who cannot, in this case, check their validity. The climate change skeptics rightly point out that temperature increase on the globe's surface is admittedly true, although it has remained at a low level — less than 1°C since the start of the Industrial Era. How, in this case, can we be assured that the leading cause is the CO_2 content increase, while many other factors such as variations in solar radiation, cosmic rays and others may also influence temperature?

In order to check this assumption, one should study the significant changes in CO_2 content and temperature in geological time scales over hundreds of millions of years. These great variations can give us an idea about the validity of the basic "greenhouse effect" argument applied to the biosphere.

But before listing the methods scientists have developed to assess concentrations of CO_2 and temperatures in these time scales, we should note the very accurate data gathered since the start of a systematic measuring in the late 1950s at the Observatory of Mauna Loa in Hawaii.

1. Photosynthesis, a powerful mechanism for the extraction of CO_2 from the atmosphere

Let us take a closer look at the measurements taken at Mauna Loa, which have shown significant variations in concentrations of CO_2 even within a period of one year. The amplitude of these variations is greater than the annual upward drift, currently 2 ppm per year. As already noted in Chapter 2, this amplitude varies considerably with latitude. It is very low at the South Pole and increases gradually when approaching the North Pole. As most of the land mass is found in the northern hemisphere, it is supposed that these variations are due to the plants and forests that cover a great part of this surface.

Figure 3.1 presents the schema of the annual cycle of variations of CO_2 concentrations at Mauna Loa (in the northern hemisphere), taking into account a drift of 2 ppm per year due to anthropogenic emissions. From April to September, the drop in concentration reaches about 7 ppm. This phenomenon is, therefore, due to photosynthesis, which is a powerful mechanism compared to variations resulting from anthropogenic emissions. In spring and summer, this mechanism extracts far more CO_2 from the atmosphere than the quantity we eject.

Nevertheless, regarding an annual cycle, photosynthesis only compensates for CO_2 emissions due to the respiration phenomenon that prevails during the winter months.

Indeed, plants and forests continue to function in winter, dipping into their reserves of stored sugars, which are burned and converted into CO_2, while the oxygen necessary for combustion is removed from the atmosphere.

Apart from the variation due to anthropogenic CO_2 emissions, the system is therefore balanced, on our scale of time, at least. The amount of CO_2 absorbed in summer by photosynthesis is equal to that ejected in winter by respiration.

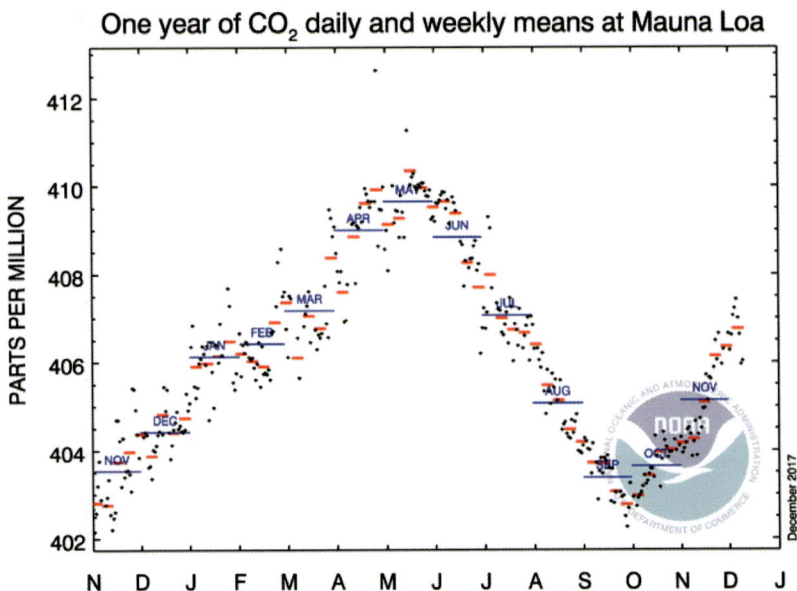

Fig. 3.1. Yearly variation of the concentration of CO_2 at Mauna Loa, taking into account the long term yearly drift of around 2 ppm/per year.

1.1. *A balance between energy and entropy*

Such maintained balance is an interesting phenomenon. We can wonder why photosynthesis manages to compensate exactly for the effect of respiration.

Obviously, if photosynthesis would not compensate for the respiration effect, plants would disappear, since they would consume more sugars in winter than they produce and store in summer.

But why doesn't the photosynthesis effect outstrip the respiration effect? In other words, why can't plants draw out more CO_2 from the atmosphere than they actually do?

As will be explained below, this is precisely what happened on geological time scales.

1.2. *When was the balance disrupted?*

The continuous increase of CO_2 concentrations as demonstrated by the measurements performed by Mauna Loa and other associated stations was a revelation. But greater revelations were the seasonal variations, which we emphasize here, as they brought to light the fragile balance of the biosphere.

The annual drift of concentration of CO_2 shows that this balance has been altered. But when did it happen? At what level of CO_2 anthropogenic emissions did the biosphere balance break down? And, therefore, what level of emissions should be recovered in order to reset the balance?

To get the right answer, we should compare the variations in CO_2 concentrations and emissions as they occurred during recent times, at the start of the Industrial Era.

In Fig. 3.2, data from the last millennium shows that this concentration remained approximately the same, 280 ppm, till about 1850, except for minimal fluctuations of no more than 3 ppm around the average value. However, as early as 1875, there was shift. At first, it was low until 1900, but then it increased rapidly. It took until 1950, about a century after the start of the Industrial Era, until concentrations of CO_2 reached 310 ppm. This value was still low compared to the 400 ppm reached in 2016.

Figure 3.3 shows carbon emissions since 1800. These emissions actually started to go up around 1850. In 1900, they are still only $0,5.10^9$ tons, about 20 times less than today. Their effect on CO_2 content, however, is already clearly visible — see Fig. 3.1.

The comparison between CO_2 content and emissions is an empiric confirmation that the biosphere balance alteration was due to the use of fossil fuels since the start of the Industrial Era. But it is only after 1950 that this break clearly deviates from the fluctuations previously observed.

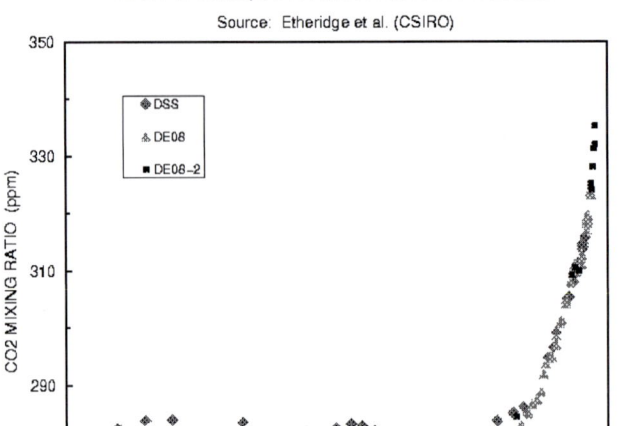

LAW DOME, ANTARCTICA ICE CORES
Source: Etheridge et al. (CSIRO)

Fig. 3.2. Variations in CO_2 atmospheric concentration during the last millenium. This concentration remains constant and equal to 280 ppm, with fluctuations of less than 3 ppm around this average value, until the end of the 19th century.

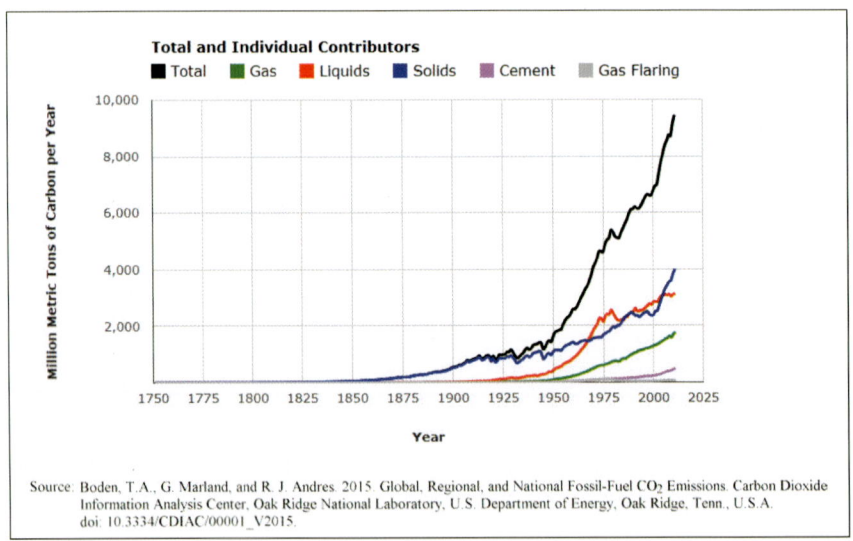

Fig. 3.3. Carbon emissions since 1800.

1.3. *The biosphere as a Carnot engine*

We may consider that the biosphere operates according to the same principles as the Carnot engine. This machine can operate in two opposite modes: the "steam engine" mode, where heat is transferred from the hot source to the cold source, work being done by the machine; or "heat pump" mode, where the work is performed from the outside and heat is transferred from the cold source to the hot source. In the first case, hot and cold are mixed, and entropy increases. In the second case, hot and cold are separated and entropy decreases.

Similarly, in winter, CO_2 molecules are transferred by the respiration process from an environment where they are in a concentrated state (a state of low entropy in the form of sugars in the plants) into an environment where they are diluted (state of high entropy in the atmosphere). Entropy increases. It is the same process as heat transfer from a hot source to a cold source performed by the Carnot engine.

In summer, CO_2 molecules undergo photosynthesis and pass from the atmosphere, where they are in a diluted state, to the plants where they are in a concentrated state (sugars). Entropy decreases. This is equivalent to the Carnot engine working in the heat pump mode.

From a thermodynamic point of view, the biosphere is in a stationary state when winter entropy increase is equally compensated by the decrease of entropy allowed by energy supply coming from outside (solar energy in this case) in summer.

This entropy decrease ΔS can be deduced from the following equation:

$$\Delta U - T\Delta S = 0$$

where ΔU represents the incident solar energy input.

In the time scale considered in Fig. 3.2, i.e. a thousand years, it can be assumed that the solar energy flow has remained constant. Consequently, the amount of CO_2 that can be extracted every year from the atmosphere, through this flow, remained the same as before 1850. Before the start of the Industrial Era, this flow was able to compensate for winter entropy increase due to the respiration of plants. But since then, this flow has been insufficient to compensate for the additional entropy increase due to CO_2 emissions by combustion of fossil fuels. The balance has been disrupted. Every year witnesses a significant increase of biosphere entropy, and a decrease of the biosphere free energy (that which is really available).

2. CO_2 atmospheric content on the million year scale

On a longer time scale, the stationary state observed up to the early Industrial Era was not the rule. Following core drilling operations, especially at the South Pole, where the oldest ice layers — exceeding 3 kilometers in thickness — are located, very accurate measurements showing changes in CO_2 content and other gases on a scale of several hundreds of thousands of years have become available.

2.1. *The core drilling method*

Snow falling on glaciers turns into ice layers that accumulate one on top of the other, and then they are gradually buried. These layers contain small bubbles of trapped air. These bubbles are air samples from the days when the ice was formed. By extracting ice cores from deep layers, we are able to access the composition of air from bygone eras. The thickest glaciers are found in Antarctica, and this enables us to know about very remote periods.

As explained in Chapter 2, the CO_2 content measured at the South Pole is a reliable reflection of its value within the atmosphere, as concentrations become homogeneous over one or two years, which is insignificant against the time scale we are focusing on. Violent volcanic eruptions might have increased CO_2 concentrations away from the South Pole on a local and temporary level, but, very quickly, the values measured there have reflected the average concentration in the atmosphere.

A drill core reaching one meter in length represents an ice deposit from around 100 years ago, which is a longer time range than the few years required for total homogenization. We can, therefore, go back to the composition of the atmosphere with a time resolution of around ten years.

This method was used to gather values as shown in Fig. 3.2, representing the past millennium. Before revealing the results of similar measurements over much longer periods, it is necessary to say a few words on the method used for dating drill cores.

2.2. *Dating of the drill cores*

The method for dating drill cores is based on seasonal variations in concentration of ^{18}O, the oxygen isotope containing two neutrons more than the majority isotope ^{16}O, found in the snow that accumulates on polar glaciers.

Both isotopes are stable, and their relative concentrations are generally constant throughout the biosphere. However, they vary with rainfall at different latitudes at a given season, and at given latitudes.

Figure 3.4 shows these seasonal variations in fresh snow, as measured during one year in Antarctica at the Dome Fuji station (Motoyama *et al.*, GPR 110 2005). The ^{18}O content is presented with the $\delta^{18}O$ parameter defined by:

$$\delta = R / <R> - 1,$$

R being the ratio between the content in ^{18}O and ^{16}O, and $<R>$ is the average value of this ratio in the biosphere.

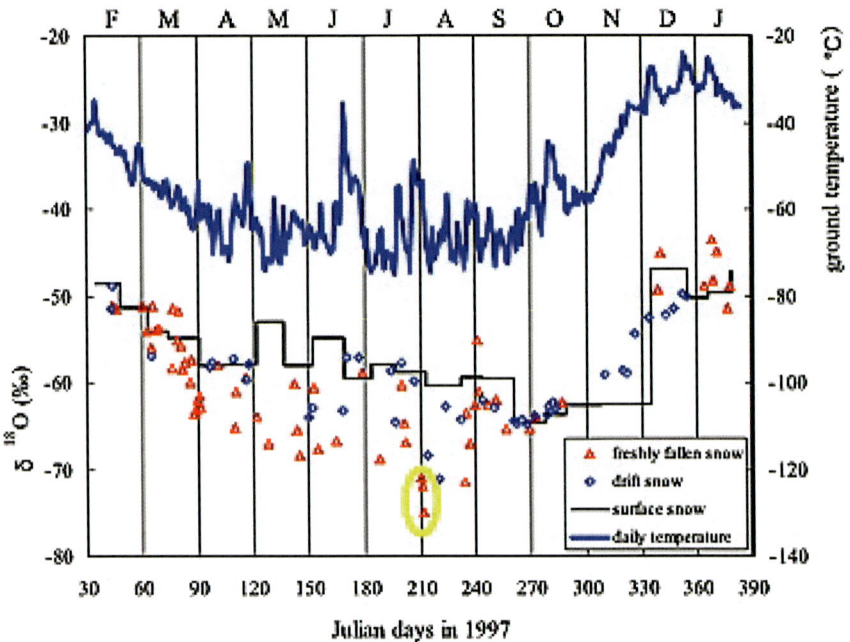

Fig. 3.4. ^{18}O content in fresh snow is lowest in (Antartica) winter.

Fresh snow is poorer in ^{18}O during winter time, when temperature is lowest.

^{18}O content decreases by around 20 ppm when the temperature goes down by 35°C. Figure 3.5 shows the correlation between ^{18}O content and temperature.

In addition, the seasonal variation of ^{18}O concentration changes with latitude, as shown in Fig. 3.6. This variation can be clearly seen in high latitudes, in the Polar Regions. As noted, the ^{18}O content is lowest in July at the South Pole, and in January at the North Pole.

Drill core dating is based on the ^{18}O seasonal variation as shown in Fig. 3.4. We can thus count the number of years elapsed for a drill core through the number of alternations between regions with strong or weak ^{18}O content. Thus, it has been possible to follow the variation of CO_2 concentrations over a period of time that goes back to 800,000 years ago.

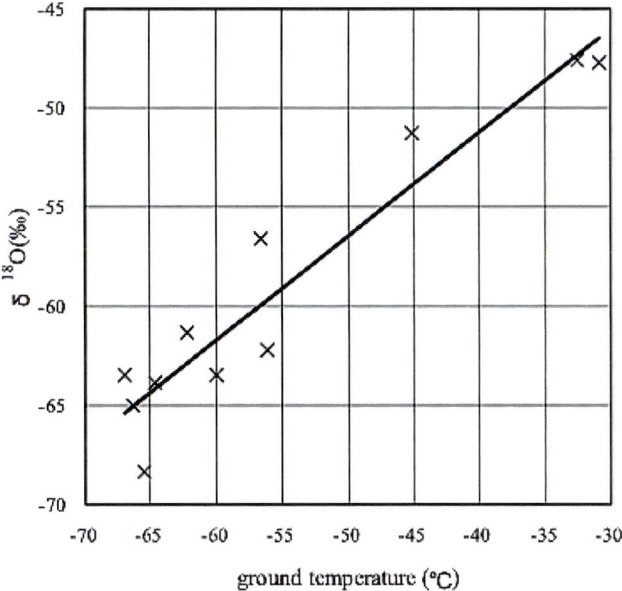

Fig. 3.5. ^{18}O content and measured temperature ratio in Dome Fuji.

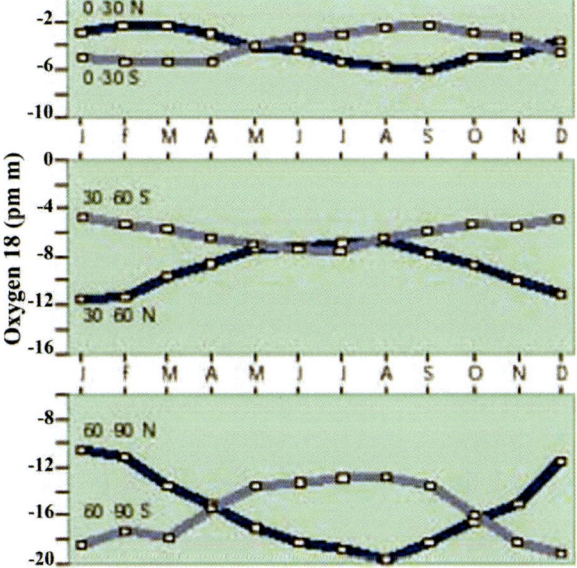

Fig. 3.6. Seasonal variation of ^{18}O content at various latitudes.

Drill cores taken at the South Pole (Epica Dome C) show significant periodic changes in CO_2 content (Fig. 3.7). Prevailing frequency reaches around 100,000 years, and the amplitude of the variations is about 100 ppm. The concentration of CO_2 varies between 180 ppm and 280 ppm for the last four cycles. Each one of these cycles begins with a rapid increase of CO_2 concentration over a period of around 10,000 years, followed by a slow decline over around 100,000 years. We are currently at the end of 10,000 years of rapid increase in a new cycle.

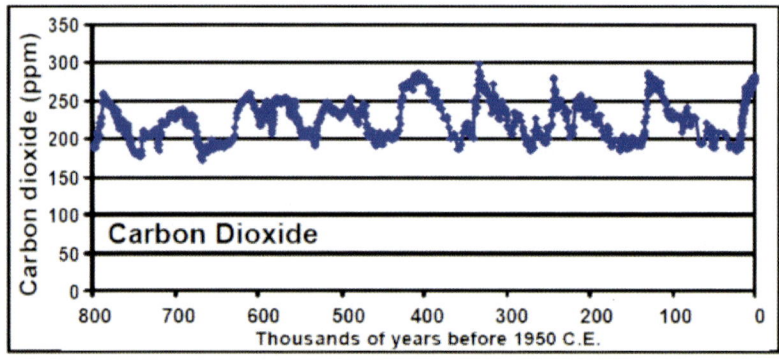

Fig. 3.7. Variation of CO_2 content obtained by the drill core method over a period of 800,000 years (Antarctic Dome C). The dominant period is 100,000 years with oscillating amplitudes ranging from 50 to 100 ppm.

As in the previous four cycles, the increase was from 180 to 280 ppm. It should be noted that during this period of relatively fast change, variations in concentrations of CO_2 did not exceed 0.01 ppm/year. This pace of "natural" variation is to be compared with the current variation, which is 2 ppm/year, or 200 times faster (as shown in Fig. 2.5d). In the biosphere history, such a quick change is unprecedented.

2.3. *Correlation between concentration of CO₂ and temperature*

While the air bubbles trapped in the drill cores enable the calculation of the concentration of CO_2 and other gases as they were preserved when the ice was formed, measuring ^{18}O isotope contents and deuterium contents in the ice enables one to calculate the temperature on the surface of Earth where and when it was formed. Figure 3.8 shows that the variation of temperatures is very similar to that of concentrations of CO_2 over 800,000 years.

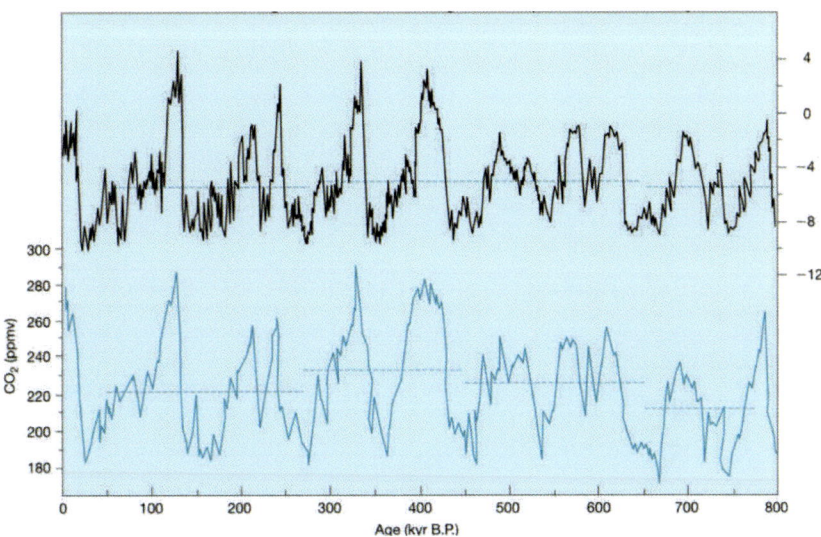

Fig. 3.8. Temperature variation follows in all its details to those of the CO_2 concentration over 800,000 years (the time arrow is reversed compared to that set in Fig. 3.7).

This figure shows that CO_2 content and temperature evolve with almost identical variations: a rapid rise over 10,000 years, before a slow decline over 100,000 years. The most recent complete cycle was characterized by a rapid rise about 130,000 years ago, with CO_2 content rising from 180 ppm to 280 ppm and the temperature increasing by almost 12°C. This fast increase was followed by an initial quick drop that became more gradual over a period of 100,000 years, before CO_2 concentrations and temperature regained their original values. The structure of these successive periods is variable, but the correlation between CO_2 content and temperature can be seen in all cases in their minutest details. For example, during the period that started 250,000 years ago and ended 130,000 years ago, we note a double peak both in CO_2 content and temperature. Another example is the exceptional duration of the warm stage with high CO_2 content for the period that began 450,000 years ago and ended 100,000 years later.

2.3. *Heavy isotope content as a temperature proxy*

We have already seen that ^{18}O content follows, on a given site, the same seasonal variation as temperature (Fig. 3.4). The relationship between these two measurements appears to be linear, at about 0.6 %/°C (Fig. 3.5).

A similar relationship between heavy isotope content and temperature has been established when comparing the measurements performed on ice samples taken simultaneously at different heights on the Antarctic continent. Temperatures differ between these various altitudes, reaching gaps of several tenths of degrees (Fig. 3.9). The use of isotopic composition as a temperature proxy is thus empirically justified.

The empirical relationship between the concentration of heavy isotopes and temperature is explained by the phenomena of evaporation and fractional condensation.

Water molecules containing one of these isotopes, deuterium or ^{18}O, are slightly heavier than those containing majority H varieties only (hydrogen) and ^{16}O. A little more energy is necessary for the heavy molecules to pass from the liquid phase to the steam phase. Water steam generated above the oceans by evaporation contains a little less of these heavy molecules than the water from which they originate. This is the phenomenon of fractional evaporation. When they rise up through the atmosphere, water molecules form clouds that are carried to the Polar Regions, where low temperatures cause condensation in the form of snowfall.

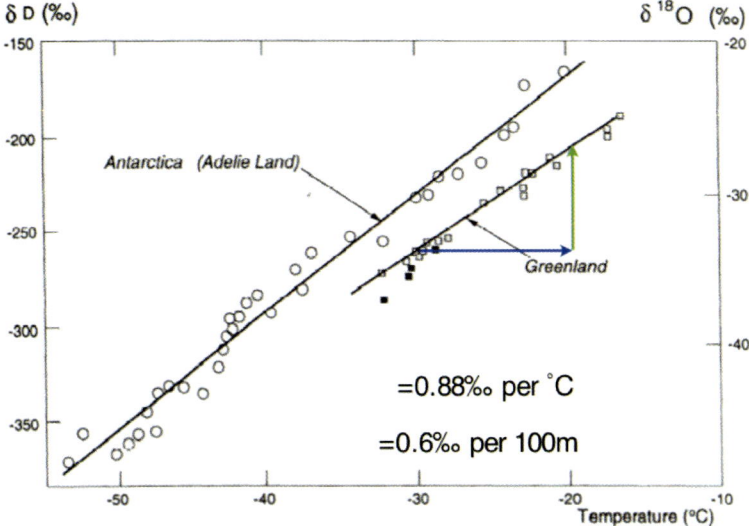

Fig. 3.9. Deuterium and ^{18}O concentrations measured in ice samples that formed simultaneously at different altitudes on the mainland of Antarctica vary linearly with local temperature.

The phenomenon of fractional condensation produces the first snowflakes that are produced over the Antarctic continent and contain a greater quantity of heavy isotopes than the vapor from which they are derived. As time goes by, as rain falls and penetrates the continent, there is a growing lack of heavy isotopes. The lower the temperature, the more pronounced is this phenomenon of fractional condensation, and therefore the snow and ice become poorer in heavy isotopes. This explains both seasonal variation (Fig. 3.6) and altitude variation (Fig. 3.9) of the concentration of heavy isotopes.

Analysis becomes more complex when trying to extend the relationship between isotopic concentration and temperature to medium and low latitudes, as can be seen in Fig. 3.6. On the other hand, if we focus on the Polar Regions only, this relationship seems to be strongly confirmed.

The close correlation between the concentration of CO_2 and temperature based on the concentration of heavy isotopes, as shown in Fig. 3.8, is spectacular. Not only do both series show the same dominant periodicity of about 100,000 years, but they have exactly the same fine structure that overlies this dominant periodicity. As these two measures are set separately, it is obvious that this fine structure is not the result of uncertain measuring. This indicates the presence of additional characteristic periods beyond the 100,000-year period.

Thus, the origin of periodic variations of CO_2 content and temperature, as well as their close correlation, is to be reconsidered. This issue is essential. We are presently at the end of a period of fast warming. Will a cooling period follow, as occurred in the previous three cycles? In that case, we are on the eve of a new ice age, and the recently noted warming should not cause us too much concern. On the other hand, if the new cycle process were to be similar to the one observed 400,000 years ago, the hot period might spread over a much longer period of around 50,000 years, and we should not rely on a future glacial period to repair the effects of our anthropogenic emissions.

2.4. *Milankovitch's cycles*

It is therefore essential to understand the origin of the succession of cold (glacial) and warm (interglacial) periods. Indeed, there is a great difference of temperature between these periods, reaching 10°C, and we are actually in a hot period that has already been going on for over 10,000 years. If it is about to end, a fall in temperature of about 8°C will follow soon — a fall that would completely upset the biosphere, resulting in glaciers extending south to latitudes similar to that of Paris.

The widely-accepted hypothesis is that periodic variations over long periods of time can only originate from astronomical factors, if we consider their regularity. This is Milankovitch's theory, which we have studied elsewhere in detail (*The Entropy Crisis*).

The basic idea is that temperature variations are due to the unstable flow of solar energy that reaches the Earth. Four astronomical factors contribute to variations of this flow.

(a) The eccentricity of the ellipse followed by planet Earth, which varies with a periodicity of 400,000 years.
(b) The angle of this orbit with the "medium" plan of the whole planet varying with a periodicity rate of 100,000 years.
(c) The angle of Earth's axis of rotation with its orbit varying every 41,000 years.
(d) The precession of Earth's axis of rotation varying at an average periodicity rate of 20,000 years.

The great accuracy with which these factors are known helps us to calculate the variation of insolation (solar radiation) at a given latitude as a function of time, and compare it with the variation of the temperature determined according to ^{18}O heavy isotope content measured in air bubbles, $\delta^{18}O$. This comparison appears in Fig. 3.10 as curves **d** and **e**. The insolation peak position often corresponds to a rise in $\delta^{18}O$ content, which marks the end of an ice age and the start of an interglacial period. This figure also shows that the rise of the $\delta^{18}O$ content often occurs simultaneously with the rise of greenhouse gas content, as seen in curves **a** and **c**. This timing was observed around 10,000 years ago, at the end of the last ice age, and at the end of the ice age that terminated about 340,000 years ago.

On the other hand, insolation does not give a clear picture of the 100,000-year periodicity that seems to dominate CO_2 concentration and temperature. It is rather the period of precession of Earth's axis of rotation, or 20,000 years, that prevails.

Therefore, the astronomical variations of insolation alone cannot explain temperature variations. Nor can they explain the periodicity variations of greenhouse gas content, which is dominated by a periodicity of 100,000 years, or the existence of the constant amplitude of these variations, passing at each cycle from 180 ppm to 280 ppm when ice age ends.

It has been suggested that this close relationship between greenhouse gas concentration and Antarctic temperature might indicate that the ocean zone surrounding the Antarctic plays a role in these variations, perhaps via a ventilation of oceanic CO_2 (Small et al., 1999).

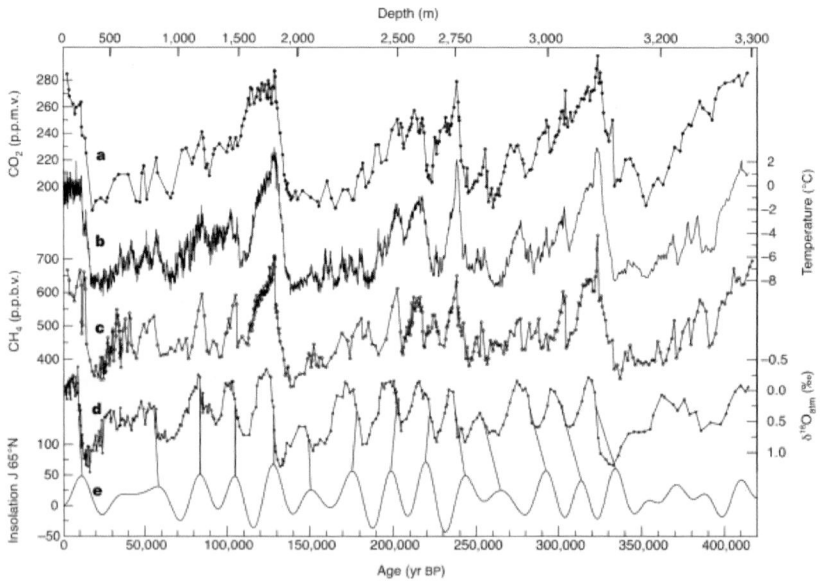

Fig. 3.10. The position of the insolation peaks (curve **e**) often corresponds to the rise of $\delta^{18}O$ (curve **d**), which indicates the end of glacial ages. But the 100,000-year periodicity that dominates changes in greenhouse gas content is not pronounced in insolation variations. Rather insulation variations are dominated by the precession of the Earth's axis of rotation, which is of 20,000 years.

More globally, it should be noted that the average amount of CO_2 in the atmosphere has not changed during the last million years.

As we have already seen, seasonal variations in CO_2 content are such that the decline recorded in summer, due to photosynthesis, is exactly compensated by the winter increase due to plant respiration. Here it is an altogether different phenomenon that controls the oscillations over a period of 100,000 years. Since the average amount of CO_2 remains constant, there is no CO_2 storage, for instance in the form of fossil fuel, but only a periodic transfer of CO_2 molecules from the atmosphere to the oceans and vice versa. At the end of each ice age, a large amount of CO_2 stored in the oceans is released into the atmosphere, just as it was about 10,000 years ago. Temperatures rise about 10°C in 1000 years. At the end of each interglacial period, CO_2 progressively returns to the oceans and the temperature drops again.

2.5. *Duration of interglacial periods*

It is generally accepted that the mechanism that triggers the end of a glacial period is a peak of astronomical insolation, as can be seen in Fig. 3.10. But there is no correlation between the amplitude of these peaks and the amplitude of the rise in CO_2 content, which remains the same during each cycle. For instance, the peak of the most recent insolation is low, but the rise in both temperature and concentration of CO_2 is high.

The peaks of insolation alone do not explain everything. The interaction between insolation and biosphere over several hundreds of thousands of years still needs to be understood.

It has been noted that the eccentricity of Earth's orbit can play a crucial role in determining the duration of interglacial periods. When it is weak, the precession of Earth's axis of rotation leads to only a slight variation of insolation. This is what happened 400,000 years ago, and this interglacial period was particularly lengthy — of the order of 50,000 years (Fig. 3.8). From an astronomical angle, we are presently in a similarly low eccentricity situation. This has lead to the view that the current interglacial period will be much longer than the previous one (Berger *et al.*, 2002, Ganopolski *et al.*, Nature 529, 200–203, 2016), and possibly of the same order as that which occurred 400,000 years ago.

The answer to the question regarding the duration of the current warm period is that it will probably be long. It is unlikely that we are on the eve of a new ice age. Thus, we should not rely on it to compensate for the warming of anthropogenic origin.

3. CO_2 concentrations on the time scale of several hundred million years

As we have seen, variations occurring in solar radiation play a crucial role in the variation of greenhouse gas concentration. This is quite clear in the case of seasonal variations, and this is also probably true regarding the scale of hundreds of thousands of years that we have been pointing at, even if the link between insolation and CO_2 content is not perfectly clear. Over these time scales, and until the start of the Industrial Era, the total amount of CO_2 in the atmosphere and the oceans has remained constant. It is the insolation variations that determine, or at least trigger, changes in CO_2 content and temperature.

On the other hand, the hundreds of millions of years scale shows a different situation. In this case, as we will see, it is the CO_2 concentration in the atmosphere that determines the variation in temperature.

3.1.   *Determination of the atmospheric CO_2 content on large time scales*

The drill core method is unable to go beyond one million years ago, which corresponds to the greatest thickness of ice in which air bubbles were trapped.

Beyond this million-year scale, dating methods rely on the analysis of fossil organisms or plants that were generated by photosynthesis. The remotest dates go back to around 500 million years. For example, phytoplankton, which forms the beginning of the ocean food chain, incorporates a fraction of ^{13}C isotopes, which depends on the concentration of CO_2 in water. The measurement of the $^{13}C / ^{12}C$ isotopic fraction thus makes it possible to go back to this concentration, which itself depends on the atmospheric concentration. Another example is that of higher plants with porous leaves that enable control of the incoming CO_2 flow. The density of these pores increases with the CO_2 content in the atmosphere, which can be found by the stomacal density method.

Figure 3.11 shows the results of indirect measurements of this type.

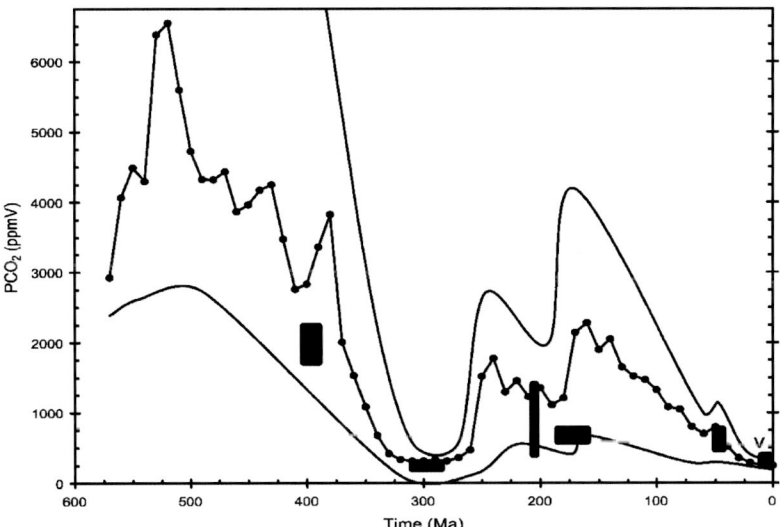

Fig. 3.11. CO_2 content obtained by the stomacal density method. The lower and upper curves represent the margin of error of the GEOCARB III model used by Berner and Kothavala to reconstruct the variation in CO_2. [According to D. Royer *et al.*, Earth Science Reviews **54**, 349 (2001).]

3.2. *CO_2 concentration variations and major glacial eras over 500 million years*

On a scale of 500 million years, the concentration of CO_2 has evolved from 8000 ppm to its current value. But this decrease has not been constant; 300 million years ago, the CO_2 content was the same as today, for about 100 million years. Another decrease, less marked, also occurred 200 million years ago.

Both periods marking the lowest level of CO_2 content — 300 million years ago, and, in recent times (on a scale of tens of millions of years) — are the major glacial eras, during which glaciers extended permanently over a large part of the northern hemisphere (Fig. 3.12c). Slight and shorter glacial periods have occurred at other times as well, but not with such a firm drop of CO_2 content.

The qualitative link between temperature and the concentration of CO_2 is, therefore, firmly established. Quantitatively, the evolution of temperature has been measured according to ^{18}O content, and corrected according to the pH evolution in the oceanic environment. See Fig. 3.12a.

As noted by Shaviv and Veizer, there is a certain similarity between the evolution of cosmic ray flux and the evolution of temperature calculated according to the ^{18}O content. These authors have suggested that it is the variations of this flux, rather than those of the CO_2 concentration, that control temperature. However, as Royer *et al.* noted, the variations of this flux do not excel in reproducing the major ice era that existed 300 million years ago, their amplitude and stages barely coinciding (it is rather low, whereas glacial ages are major). On the other hand, calculated temperatures, when taking into account the pH variation, perfectly reproduce CO_2 content variations as shown in Fig. 3.11.

Thus, it is obvious that on a scale of hundreds of millions of years, variations in atmospheric CO_2 content, rather than the variations of the insolation or the flux of cosmic rays, have determined changes in temperature. Over this very long period of time, it is the decrease in CO_2 content that has caused a decrease in temperature.

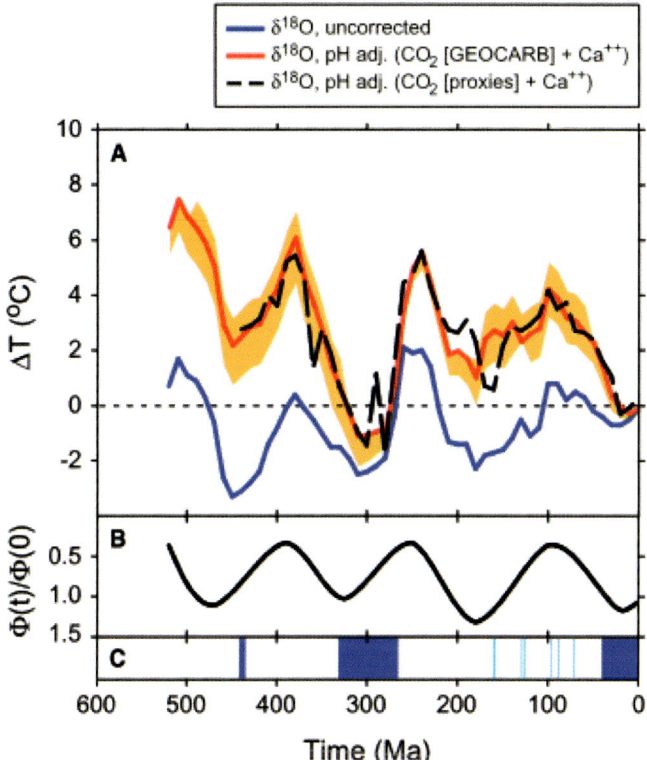

Fig. 3.12. From top to bottom in A: change of the temperature evaluated on the basis of the ^{18}O isotope content, taking into account the evolution of the pH, compared to the GEOCARB model; uncorrected evolution; B flux of cosmic rays; C glacial eras according to geological surveys (after Royer *et al.*, Op. cit.).

4. Origin of the CO_2 concentration variations on the hundred million years time scale

Over a period of 500 million years, the major characteristic change in atmospheric CO_2 content is a decrease of around a factor of 20. It changed from 8,000 ppm to 400 ppm. CO_2 removed from the atmosphere has been stored in the form of carbonates and fossil fuels, the basic mechanism, in both cases, being photosynthesis. But as can be seen in Fig. 3.11, this storage was not created gradually.

Over a relatively short time of 50 million years, from −325 million years to −375 million years, CO_2 concentrations passed from approximately 4000 ppm to a value close to that at present times. It remained at this level for about 100 million years. As Fig. 3.12 shows, the decline in CO_2 content started before a major ice age. This confirms that it was the drop in the CO_2 content that led to the fall in temperatures.

4.1. *The Carboniferous Period*

During the sharp decline of CO_2 content, and at the start of the ice age that followed, a large part of our fossil fuel reserves was formed. This period is called the Carboniferous Period. It ranges from −350 to −300 million years. It was then outstripping geological conditions, rather than a particularly strong insolation, that were the cause of this rapid storage. Landmasses were beginning to cluster and form a single continent called Pangea, which was located in the southern hemisphere. High temperatures and heavy rains enhanced plant growth. The separation between dry lands and marshy areas was not as clear as it is today. Dying plants, instead of decomposing on the mainland, were transformed into peat in these swamps. Over geological time scales, these carbon-rich layers were buried and compressed, and transformed into charcoal and other fuels before surfacing far from their formation places, during powerful tectonic events.

4.2. *Estimates of fossil fuel reserves*

We can evaluate coal reserves if we accept that they were mainly produced during the Carboniferous Period, when CO_2 content had decreased by about 4000 ppm. During this period, storage in the form of coal was undoubtedly the prevailing form over the formation of carbonates, which is a slow and regular process.

If we compare the pace of the current increase in CO_2, i.e. 2 ppm/year, to that of carbon emissions, which is 8.10^{11} gC/year, the result is that about half of the emissions are currently dissolved in the oceans. If all emissions were stored in the atmosphere, CO_2 content would increase by 4 ppm/year. Therefore, the carbon contained in all fossil fuels stored during the Carboniferous Period — coal, oil and gas — would return to the atmosphere within 1000 years. Taking into account the pace of their exploitation, and assuming that all these reserves are workable, the temperature at the end of this period would be a dozen degrees higher than it is today. The rate of increase in temperature would be about a degree per century, which corresponds quite well to the measured increase since the start of the Industrial Era.

Even if we were on the eve of a new ice age, these effects would not be felt at the end of 1000 years.

4.3. *Change rate of CO_2 atmospheric content in geological times and in present times*

In the most favorable conditions that ever existed, say, during the Carboniferous Period, 4000 ppm atmospheric CO_2 was stored as fossil fuels over a period of 50 million years. This storage rate is 0.0001 ppm/year. It is 40,000 times slower than the pace of current emissions. This comparison shows the extent to which the biosphere is disrupted today. Over the timescale of our civilization, we cannot rely on photosynthesis to restore the balance.

This conclusion is consistent with the study of seasonal variations in CO_2 content.

We have seen that, with the effect of photosynthesis, CO_2 content can decrease by several ppm over a period of some months. This decrease is parallel to the emissions rhythm, which implies that photosynthesis can, in fact, quickly restore the balance of the biosphere. But, in reality, it is not so. Respiration of plants tends to cancel out the effect of photosynthesis. In addition, carbon storage in the long run requires very specific conditions that did not prevail in the Carboniferous Period. We cannot count on photosynthesis to restore the balance. Even an extension of forests would not suffice.

As we have seen, the link between CO_2 concentration and temperature over geological time scales, has been firmly established. At the current rate of emission, over a period of only 300 years, the CO_2 concentration would recover the value it bore 100 million years ago, or 1000 ppm. According to Royer *et al.* (Fig. 3.12), temperature was 4°C higher than it is today. This experimental data is close to the results of current models, which foresee an increase of 4°C in 2100 if CO_2 content reaches 900 ppm (see Figs. 2.2 and 2.3). Provided a 2-factor, geological data based on CO_2 content measurements and on the evaluation of temperatures defined by the isotopic method on the one hand, and the results of calculation models on the other hand, correlate.

Whatever the method of calculation used, an increase in temperatures of the order of several degrees by 2100 seems inevitable if we continue to burn fossil fuels at the current rate. And even if we are on the eve of a new ice age, this would not change anything, because it would only have effect over a much longer period of time.

4

Thermodynamic Conditions for a Return to Equilibrium

As shown in detail in Chapter 3, the increasing rate of increase of CO_2 content in the atmosphere during recent decades is an exceptional situation in the history of the biosphere.

Figure 3.8 presents how, at the end of each glacial period, the concentration of CO_2 has risen by around 100 ppm over 10,000 years, which means an increase of 1 ppm per century. But during the last century, it has increased by 100 ppm. In recent years, it has even increased at a rate of 2 ppm per year, or 200 ppm per century. This is 20,000 times faster than the rate of increase measured at the end of the last glacial period. It is truly an upheaval for the biosphere.

This rate can also be compared to the capture rate of CO_2 by photosynthesis on geological time scales. Capture rate is 100,000 times slower than the emission rate measured over recent decades. We burn fossil fuel reserves 100,000 times faster than their rate of accumulation.

The very consequences of this extremely rapid concentration increase of CO_2 are difficult to predict, but it is clear that the biosphere has entered a new, unknown cycle. There are signs of instability everywhere, as mentioned in the COP 21 Agreements.

Therefore, returning to equilibium as fast as possible — that is, returning to a CO_2 concentration of approximately 300 ppm — is highly desirable.

Economic and especially political means that are required for this recovery have been widely discussed and negotiated as part of the agreements signed by the COP 21 Parties. They call for measures that are to various degrees restrictive. They are different from those suggested by Malthus, but have in common the understanding that a dreadful catastrophe cannot be prevented by regular economic factors, i.e. mechanisms of a market economy. Nevertheless, there is a price involved in these agreements that seek a recovery of the balance of the biosphere. It should, therefore, be estimated.

In this chapter, we recommend the use of the laws of thermodynamics, and, more specifically, the notion of entropy introduced by Carnot and established by Clausius, to evaluate this price. We will show that temperature increase and other signs of climate disturbance, such as air and water pollution, as well as agricultural soil deterioration, are all a result of the biosphere's increased entropy. This increase is due to the inconsiderate use of natural resources, fossil fuels being just one example among many others. This use leads to the accumulation of waste of all sorts. What is missing most, rather than oil or coal, is the empty space to put all this waste in.

Thus recycling this waste is a necessity. It is generally possible, but this implies costly energy input. It is not surprising. As stressed by Carnot, Clausius and Boltzmann, reducing the entropy of a system is possible, provided there is a supply of external energy.

We will focus on the case of greenhouse gas contribution, mainly CO_2. The CO_2 emissions increase the entropy of the atmosphere, and thus create pollution, even if CO_2 molecules by themselves do not have any harmful effect. There are other forms of pollution that are directly harmful, such as fine particles resulting from the operation of internal combustion engines; they might become even more harmful in the near future. However, the example of CO_2 is straightforward to calculate.

It is, therefore, necessary to estimate the increase of entropy in the biosphere resulting from the emissions of this gas, and further, to evaluate the energy that would be necessary for its capture, based on the principles of thermodynamics.

To help us answer this question let us return to Sadi Carnot's ideas that were introduced in Chapter 2.

1. The biosphere as an entropic Carnot machine

Our calculation relies on the same principles that govern the operation of the Carnot engine. Let us thus go back to these principles.

1.1. *The "caloric" Carnot heat engine*

Carnot describes his model engine as follows:

> There are two bodies, A and B, each kept at a constant temperature, that of A being higher than that of B; these two bodies to which we can give, or from which we can remove heat without causing their temperatures to vary, exercise the functions of two unlimited reservoirs of caloric. We will call the first one the furnace, and the second one the refrigerant.

At the time of Carnot, father and son, heat flow is considered to consist of particles known as "caloric." In fact, such particles do not exist. The only particles we have are the molecules that compose fluids (here, water vapor, or liquid water), as Boltzmann will later demonstrate. These molecules have a greater kinetic energy in hot fluid than in cold fluid. When a molecule goes from the hot reservoir, A, to the cold reservoir, B, it transmits some of its kinetic energy from A to B. This is how a quantity of heat, Q, passes from the furnace to the refrigerant. Calorics are rather insignificant in Carnot's theory: their role is for illustrative purpose only. They do not alter the generality of his demonstration, which supposes that both bodies, A and B, remain at constant temperature, regardless of the quantity of provided or extracted heat.

The Carnot model engine can work in two different modes: the thermal machine mode (steam engine) when a quantity of heat, Q, passes from furnace A to refrigerant B, and work W is provided by the machine to the outside; and the heat pump mode when work W is provided *from the outside*, and a quantity of heat, Q, is transferred from refrigerant B to furnace A (Fig. 4.1).

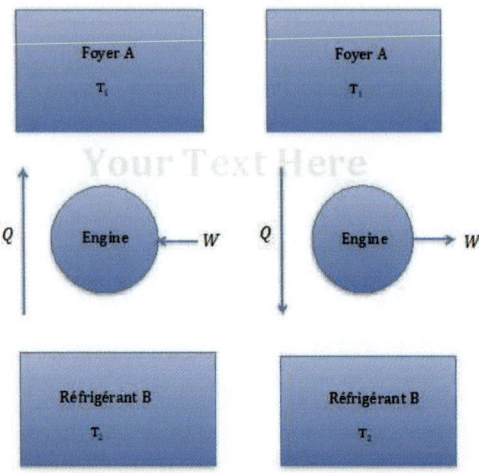

Fig. 4.1. The Carnot engine operating in a heat pump mode (left) and in steam engine mode (right). In the heat pump mode, the outside world provides work, W, and a quantity of heat, Q, goes from the cold source (the refrigerant B) to the hot source (the foyer A). In steam engine mode (the one studied by Carnot), a quantity of heat, Q, passes from the foyer to the refrigerant, and the machine provides work, W, to the outside world.

Carnot described the operation of his engine as an ideal cycle consisting of four phases. This particular cycle, which bears his name, is made possible by the fact that the temperatures of the furnace and of the refrigerant remain constant during the operation of the machine.

Figure 4.2 is a schematic representation of the Carnot cycle.

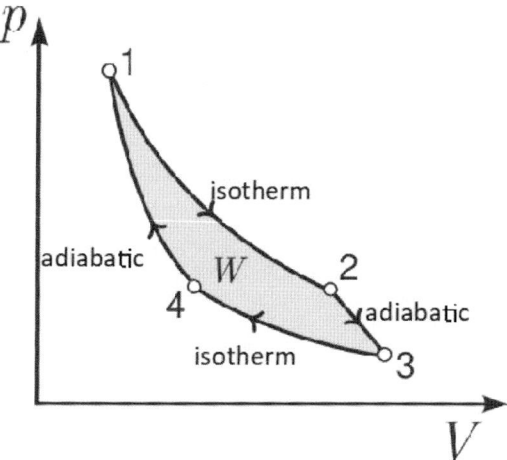

Fig. 4.2. The Carnot cycle consists of two isothermal branches (heat exchange at constant temperature) and two adiabatic branches (without heat exchange). When run in the clockwise mode of the cycle as shown in this diagram, the machine provides work, W, to the outside world.

(i)   From point 1 to point 2, furnace A transfers a quantity of heat, Q, at a constant temperature, T_1, to a fluid under high pressure (for example, steam), which moves a piston in a cylinder. *The ensuing pressure/volume curve is called an isotherm. In this branch of the diagram, the pressure in the cylinder decreases, and the volume occupied by the gas increases.*

(ii) From point 2 to point 3, the fluid is cooled to temperature T_2 of refrigerant B without heat exchange. *This branch is called adiabatic.*

(iii) From point 3 to point 4, the fluid is compressed while remaining at the temperature of the refrigerant, T_2, restoring the same amount of heat as received during the first phase *(this branch is isothermal).*

(iv) In the fourth phase, from point 4 to point 1, the fluid recovers the furnace temperature without heat exchange *(adiabatic).*

In total, the amount of heat Q, supplied to the fluid by the furnace is transmitted to the refrigerant, and work W is supplied to the outside world by the movement of the piston. Indeed, the work done during the expansion phase of 1 to 2 is greater than the work performed during the compression phase, because the pressure is higher during the expansion phase while the movement of the piston remains the same.

The problem faced by Sadi Carnot was the calculation of performed work. In his engine model, only three parameters were involved: the furnace temperature, the refrigerant temperature, and the amount of heat that passes from one to the other.

Carnot argued that performed work is proportional to this amount of heat and to the *difference* between the two temperatures. This is the simplest hypothesis, since the work provided will be zero if no heat passes from the furnace to the refrigerator, and if they are both at the same temperature.

So:

$$W = Q \, (T_1 - T_2) f(T_1, T_2)$$

where $f(T_1, T_2)$ is an unknown function of the furnace and refrigerant temperatures.

1.2. *The efficiency of the Carnot engine and Clausius' entropy*

According to a further definition stated by Clausius, the entropy variation of a system is equal to the heat input divided by the temperature of the system. In a Carnot cycle, the entropy of the system varies only during isothermal phases 1 and 3, since, in the adiabatic phases, 2 and 4, there is no exchange of heat. As the furnace temperature is the highest, in absolute terms, the entropy variation Q/T_1 in phase 1 is lower than in phase 3, Q/T_2. When the cycle is described in a 1234 order, it means that the engine has provided work outside, and that the entropy of the system

$-Q(1/T_2 - 1/T_1)$ has increased (obtained heat is counted as negative). As long as temperatures of the furnace and the refrigerator are close and almost equal to their average value, T, the change of entropy in steam engine mode becomes:

$$\Delta S = Q(\Delta T/T^2)$$

where

$$\Delta T = T_1 - T_2$$

If the operation of this engine is optimal, that is to say, reversible, free energy remains constant, and the quantity $T\Delta S$ remains equal to the supplied work W:

$$W = Q(\Delta T/T)$$

This is exactly the result obtained by Sadi Carnot when function $f(T_1, T_2)$ is simply equal to $1/T$.

1.3. *The biosphere as an entropic Carnot engine*

When the Carnot engine operates in steam engine mode whilst supplying energy, entropy increases. Entropy decreases when it runs upside down in heat pump mode. The cycle is then described in the order 4321. The work received from the outside brings back the refrigerant heat to the furnace.

A similar analysis can be applied to the functioning of the biosphere. It contains four essential elements: plants, the atmosphere, water and CO_2, with an external source of energy (the sun).

During the day and during summer, energy provided by photons emitted by the sun allows plants to combine water molecules and atmospheric CO_2 to produce sugars, the simplest being glucose, $C_6H_{12}O_6$ (Fig. 4.3). This is the so-called phenomenon of photosynthesis. Six molecules of CO_2, and six molecules of water, H_2O, form a molecule of glucose, while six molecules of oxygen are released into the atmosphere. Entropy decreases as a total of twelve free molecules have been combined to produce a complex fixed molecule (glucose) and six free molecules (oxygen). In the heat pump mode of the Carnot engine, the work provided from the outside makes it possible to bring heat from the refrigerant to the furnace. In the biosphere, photosynthesis pushes up the molecules from a state of high entropy and low energy (in the atmosphere) to a state of low entropy and high energy in plants. We could call this the entropy pump mode of the biosphere.

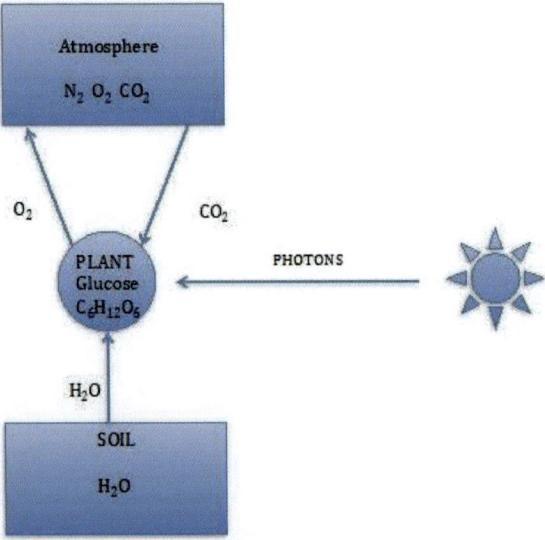

Fig. 4.3. Under solar radiation plants combine CO_2 molecules extracted from the atmosphere with water molecules pumped from the soil to produce glucose, $C_6H_{12}O_6$. The carbon atoms move from a state of high entropy (incorporated into free CO_2 molecules in the atmosphere) to a state of low entropy (incorporated in the glucose found in the plant). In the absence of solar radiation, the plant burns its glucose to survive, the carbon returns to a state of high entropy in the atmosphere.

At night, and during winter, plants burn their sugars for survival by using oxygen from the atmosphere, while emitting CO_2. By analogy with the Carnot engine, sugars are the furnace, while the atmosphere becomes the refrigerant. The biosphere operates in thermal machine mode. Entropy increases because, when released into the atmosphere, the CO_2 molecules can explore a much larger number of sites than if they are incorporated into the glucose molecules.

1.4. *The annual cycle of CO_2 as an illustration of the functioning of the biosphere*

The CO_2 annual cycle measured at Mauna Loa and at other meteorological stations reflects these phenomena on a global scale. The entropy pump effect is powerful. Figure 4.4 shows an average of these measures for the period 2012–2016 as recorded in different parts of the world. During this period, concentrations of CO_2 increased by 2.2 ppm per year. Based on these data, annual cycles appear clearly. CO_2 concentrations increase during three quarters (autumn, winter and spring) up to a total of 7 ppm, during which plants burn more sugars than they store. The entropy of the biosphere increases. During summer, the concentration decreases by 5 ppm as plants replenish their sugar reserves through photosynthesis. The entropy of the biosphere decreases. The amplitude of the cycle is significant. Without the multi-annual shift, this amplitude reaches a value of 6 ppm.

This cycle mainly reflects the behavior of forests. A relevant question would be why, each winter, do they burn exactly the amount of sugar they accumulate in summer, and this at all latitudes.

It is not only the average behavior that is worth looking at, but also the local one — see Figs. 2.5a and 2.5b. Another surprise is that the amplitude of the cycle is higher in northern latitudes, where it reaches nearly 10 ppm (Fig. 2.5a), yet it is not there that sunshine is the strongest. Hibernation clearly plays a role in the CO_2 cycle.

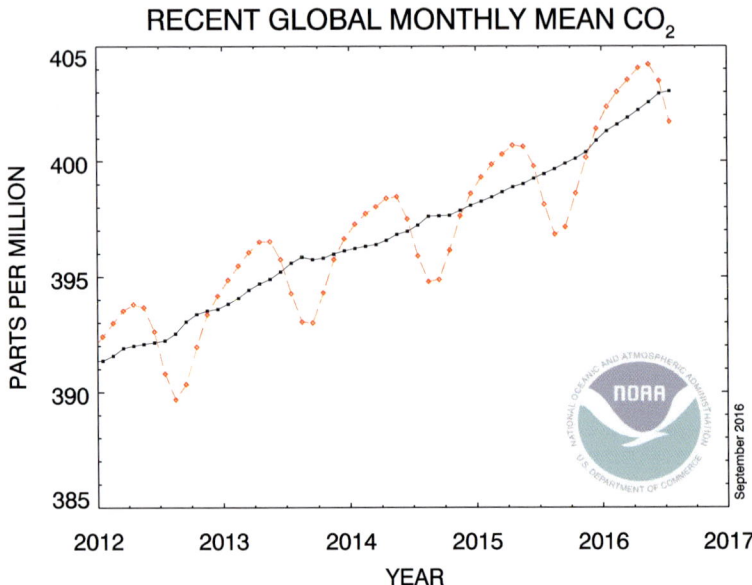

Fig. 4.4. Average annual cycles of CO_2 concentrations at several measurement stations in different parts of the world. Superimposed on the multi-annual shift of 2.2 ppm/year due to anthropogenic emissions, these cycles show an amplitude of 6 ppm.

2. The Carnot engine and the effect of environmental finitude

The invention of the concept of hot source and cold source at constant temperatures is undoubtedly a hallmark of Carnot's genius: it is this very concept that led him to conclude that the maximal efficiency of a thermal machine depends only on the amount of heat exchanged between hot source and cold source, and their temperatures. He emphasized that the nature of the fluid used and the details of the machine play no role in the value of this maximum yield.

For Carnot, at the time, the increase of this yield was the main goal in order to get maximal work out of the coal burned. The environment was not an issue, then. But did it have an impact on Carnot engine? No. It was actually eliminated by the invention of the brilliant concept of constant temperature sources, and therefore supposed to be infinite. But such sources do not exist, precisely because our environment is *finite*. We find, here, in a precise mathematical form, the cardinal importance of the finitude of the world in which we live, which is the biosphere.

Indeed, Carnot engine operation shows us that, as hot and cold sources are actually finite, when heat passes from the hot source to the cold source, the first one cools a little and the second one warms a little. After the performance of many cycles, the difference in temperature between both sources decreases gradually, and the maximum theoretical yield decreases, approaching eventually the limit of zero. This is not surprising. If we include the Carnot engine and the *finite* hot and cold sources in a same closed system, entropy must increase with each cycle, the temperatures of both sources must equalize, and the engine will not provide further work. More precisely, an increasing amount of coal will have to be burned to produce the same level of work. The next stage will be a collapse, originating less from the finitude of resources, but rather as a result of the increase of entropy itself, caused by environmental finitude.

We will now present the estimated external energy input that would be necessary to ensure a return of the biosphere to a state of balance.

3.   Energy and entropy in the biosphere

As a consequence of Greenhouse gas emissions their atmospheric concentration has increased. It is precisely this increase that is considered to be the origin of global warming, and, more generally, an essential factor in climate disturbance. Their capture, especially CO_2 capture, has become a worldwide challenge.

From a thermodynamic point of view, the increase in concentrations of CO_2 translates into an increase of entropy of the biosphere. In this section, we propose to calculate the amount of energy necessary to compensate for this entropy increase, so as to return to the previous status quo. At a later stage, we will see how this energy compares to the energy obtained from the sun, our only external source of energy. We will then use the measures shown in Fig. 4.3 to compare between the amount of CO_2 actually extracted from the atmosphere by photosynthesis and the related solar energy.

3.1. *The stationary thermodynamic state of the biosphere*

A stationary thermodynamic state of the biosphere is characterized by free energy that remains constant, despite variations of entropy and energy.

In a steady state, any increase in biosphere entropy must be compensated for by an external supply of energy. Entropy increase in the biosphere due to the emission of a free molecule, Δs, is given by Boltzmann's Law:

$$\Delta s = k_B \ln N$$

where k_B is Boltzmann's constant and N is the number of sites that this molecule can explore. Our first task is, therefore, to estimate the number of sites.

3.2. *The diffusion process and Boltzmann's Law*

In order to evaluate the number of sites that a CO_2 molecule can explore, we must return to the mechanism that leads to irreversibility, namely to the irreversible decrease of free energy associated with the diffusion of this molecule. This irreversibility is not the result of one molecule's behavior, but that of a large number of molecules. If a large number of CO_2 molecules is released at the same time at a given point, they will gradually scatter into all the space that is offered to them, and end up in a uniform distribution. Chances that they will be gathering back together decrease as they get more numerous, and as the space offered to them is larger. If we wait long enough, this will happen, but only on a timescale that would soon surpass the age of the universe. This is the meaning of irreversibility.

The diffusion of a CO_2 molecule in the atmosphere results from collisions with the other molecules that compose it, mainly N_2 and O_2. Without these molecules that constitute the atmosphere, CO_2 molecules would follow the planned trajectories set by Newton's Laws of Mechanics, trajectories that are preserved by time reversal; there would be no irreversibility. In the presence of the molecules that compose the atmosphere, CO_2 molecules undergo a series of collisions that modify their trajectories. Irreversibility appears.

Therefore, the collisions between the CO_2 molecules and N_2 and O_2 molecules will determine the number of sites explored by a CO_2 molecule, rather than the dimension of the space. Supposing that the number, N, of different sites that can be explored by a CO_2 molecule is equal to the number of molecules with which it can collide, this same number will be equal to the number of molecules that form the atmosphere. For evaluation, we divide the weight of the atmosphere 5.10^{18} kg by the average weight of the molecules that compose it, mainly N_2 and O_2. This average weight is 28.8 g per molecule, divided by the Avogadro number. In total, we obtain:

$$N \cong 1.10^{44}$$

The entropy increase per emitted molecule is therefore approximately:

$$\Delta s = 100 \, k_B$$

3.3. *Extraction energy*

To compensate for this entropy increase and ensure a stationary state, it is necessary to provide energy per emitted molecule as follows:

$$\Delta u = T\Delta s$$

The obtained result at room temperature, $T = 300K$ is,

$$\Delta u \cong 5.10^{-19} \text{ joules}$$

In order to reduce the concentration of CO_2 (or of any other polluting molecule) by one part per million (1 ppm), one needs to extract 10^{38} molecules from the atmosphere. This will require an energy input:

$$U/ppm \cong 5.10^{19} \text{ joules}$$

This is the energy that must be provided to bring back the free energy system to its initial value, that is, the value set before the introduction of these 10^{38} molecules. It is, therefore, equal to the minimum energy that is required for the extraction of these molecules.

At the present time, total CO_2 emissions would increase its atmospheric concentration by 5 ppm per year if it were not for ocean storage. To collect back these molecules one would need an energy input of $2.5.\ 10^{20}$ joules. To collect back CO_2 that actually accumulate in the atmosphere — 2 ppm per year at the current rate — one would need an energy input of $1.\ 10^{20}$ joules. This is a considerable amount of energy extraction.

We can make a first comparison with the world consumption of energy from all sources. According to the International Energy Agency, this consumption was of 6.10^{20} joules in 2012. In other words, total world energy consumption currently is not much larger than the minimum energy needed for the extraction of emitted CO_2.

In order to reduce CO_2 concentration from its current value of 400 ppm to its "natural" value of 300 ppm, it would be necessary to devote an amount of energy ten times higher than the total annual energy consumption. This situation is obviously untenable.

We can also compare the extraction energy with the heat of carbon combustion, which is 33 MJ/kg. The extraction energy is 25 kJ/kg in these same units. The extraction of a carbon atom from the atmosphere needs an amount of energy similar to the one obtained by burning this atom. In other words, a power station supplying electricity by burning coal should use all the produced energy to capture emitted carbon in the form of CO_2. To store it back in its original carbon form would require even more energy. This is not surprising. Otherwise, we would deal with a carbon recycling that allows burning and then starting all over again.

3.3. *CO_2 extraction by photosynthesis*

The above estimates are a lower bound for the extraction energy of a pollutant molecule of any kind. Here we will focus on the case of CO_2 molecules, even if other molecules such as methane also play a role. One could conceive realistic mechanisms for the extraction and storage of CO_2 emitted during coal or hydrocarbon combustion, but only if they are performed before CO_2 is released into the atmosphere, that is, when the entropic change remains weak. But as soon as diffusion throughout the atmosphere has occurred, photosynthesis appears to be the only practical way to ensure CO_2 capture.

3.3.1. Incident solar energy and photosynthesis conversion

At first glance, the amount of solar energy received by the planet seems to be of the right order of magnitude for the quick capture of all emitted CO_2.

Forests occupy about 10% of the Earth's surface, i.e. 4.10^{13} m². The average solar flux incident over the year amounts to 150 W/m². There are about 3.10^7 seconds in a year. This gives an incident energy of 2.10^{23} joules/year.

But it takes 8 photons having an energy higher than 2eV to have a carbon atom moving from a CO_2 molecular state to a glucose state. The average energy of photons of the solar spectrum being of 2eV, quantum efficiency is of the order of 10%. The incident energy exploitable for photosynthesis is, therefore, only 10% of the incident energy, or 2.10^{22} joules/year. This energy is 500 times higher than the energy required for the extraction of 1 ppm of CO_2 per year. At first glance, the available solar energy would be sufficient to capture the excess of CO_2 in the atmosphere within one year.

But, in fact, during one year, the amount of CO_2 recycled during summer by glucose photosynthesis is burned in winter. Our calculation is correct, but it does not take into account the respiration effect. It renders photosynthesis inoperative as a rapid means of capture and storage of excess atmospheric carbon, at least in our time scale.

Nevertheless, the effect of photosynthesis is clearly visible in the CO_2 cycle. The amplitude of this cycle, especially in northern regions (Fig. 2.5a) is such that, within six months, the concentration of CO_2 decreases at an annual pace of 10 ppm, or 20 ppm on a yearly basis.

3.3.2. Effective carbon storage by forest production

Storage by photosynthesis, even when taking into account the respiration process, is nevertheless not totally ineffective. Calculations based on world annual wood production allow us to obtain a rather good estimate of its global outcome.

According to statistics published by the FAO in 2014, stored wood (timber, wood for pulp and others) amounted to $1,8.10^9$ m^3, with a similar amount for firewood. Carbon content is about half the weight of wood. If we retain only stored wood, the forests have removed from the atmosphere about 1 Gt of carbon per year by photosynthesis, or about 10% of emissions from fossil fuels. If we include firewood as a source of green energy (since it is a renewable energy source) the removed amount is of 2 Gt of carbon, or 20% of the emissions that have been compensated for thanks to the energy provided by the sun, or 0.4 ppm of CO_2 compared to the 2 ppm added each year.

Table 4.1 summarizes some of the carbon data in the biosphere.

Table 4.1. Biosphere carbon data, in gigatons of carbon per year.

Emissions	Stored Wood	Burnt Wood	Atmosphere	Oceans
10 Gt	1 Gt	1 Gt	4 Gt	5 Gt

The evaluation of the amount of carbon stored in the oceans is based on the difference between, on the one hand, emissions, and on the other hand, wood produced by photosynthesis and stored as lumber, pulp and other products, plus the annual increase of CO_2 concentrations in the atmosphere. The annual increase of 2 ppm translates into 4 Gt of carbon. Wood produced by photosynthesis and burned (as firewood or for cooking) does not change the carbon balance.

4. Balance disruption

The solar flux on the surfaces occupied by forests is by two orders of magnitude higher than the minimum energy input required to capture emitted CO_2 and to ensure the stability of the biosphere with constant free energy. The current imbalance results from the weak effective efficiency of photosynthesis, which is of the order of 1 per 1000. This low net yield from photosynthesis is mainly due to the phenomenon of respiration and decomposition of plants on Earth, and similar phenomena occurring in the oceans environment, which reduce by two orders of magnitude the quantum efficiency of photosynthesis.

The limited atmospheric carbon storage by photosynthesis is further reduced by the fact that half of the carbon produced is immediately burned and re-emitted. The geographical distribution of this use of carbon produced by photosynthesis shows a considerable difference between Africa and Asia, and other parts of the globe.

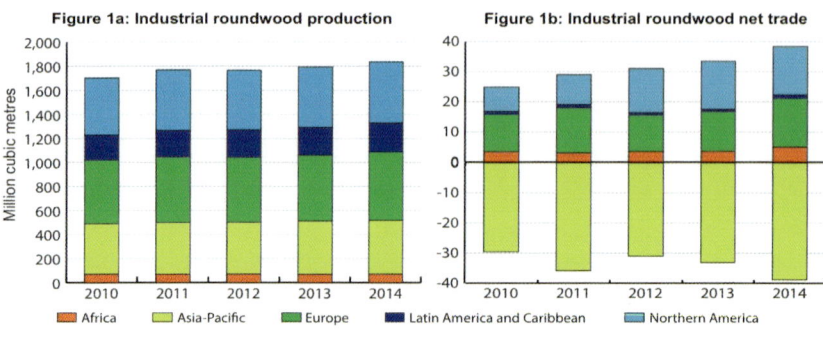

(a)

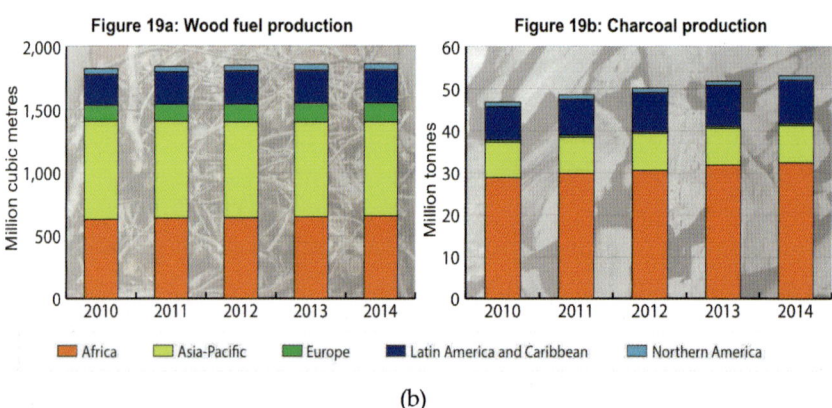

(b)

Fig. 4.5. Production of Construction Wood (a) Production of Combustion Wood (b) (FAO Forest Products Statistics 2014).

As shown in Fig. 4.5, wood is used as fuel mainly in Africa and Asia, probably for cooking food rather than heating, since winters are mild in these regions. This is a poor use of solar energy. Indeed, the photosynthesis yield is much weaker than that we could achieve from very simple solar ovens. Burnt wood obtained by photosynthesis to cook food (or heat homes) is a thermodynamic aberration that further contributes to deforestation. A solar oven with a capture area of 1 m^2 provides the same energy as 100 m^2 of forest. A massive effort to introduce solar ovens in Africa and Asia could significantly improve the biosphere carbon balance since 1 Gt of additional carbon could be removed from the atmosphere and stored every year.

4.1. *Production and long-term storage of carbon obtained by photosynthesis*

The amount of carbon produced and stored annually, or 1 Gt, is low compared to emissions of 10 Gt, but is 10,000 times higher than the carbon amount stored annually in geological time scales!

Indeed, carbon storage in the form of wood produced by photosynthesis is not a long-term storage. Over a century, this wood is destroyed, either burned or decomposed, and the carbon is re-emitted as CO_2.

Long-term storage of carbon produced by photosynthesis requires keeping wood away from fire, and from oxidation, which leads to its decomposition. This is what happened over geological time scales, when mainland and humid areas were less separate than they are today. Dying trees were then progressively buried in the swamps, as still occurs in marshy areas, and were gradually transformed into lignite and more elaborate forms of coal.

Today, most of the forests exploited for production of wood are located in dry areas. This is particularly the case in northern areas such as Canada, Scandinavia, and Russia, where a lot of wood is produced. Unfortunately, the carbon it contains cannot be stored on long-term basis. But over a century, it can contribute and take part in the desired transition to a low carbon economy.

4.2. *Acceptable limit of CO_2 emissions*

Data that we have shown above suggest that an acceptable level of CO_2 emissions should be around 1 to 2 Gt of carbon per year. This assessment is based on the following observations:

(i) This is the amount of wood annually produced by photosynthesis, see Table 4.1. This wood could be stored, and the carbon that it contains would be then removed from the atmosphere.

(ii) This was the level of emissions before the oil era that began in 1950. The rapid increase of atmospheric CO_2 has occurred simultaneously with the increase of emissions after this date. The increase in temperatures then occurs with a delay of about twenty years.

There are, in principle, ways other than photosynthesis to remove CO_2 from the atmosphere. But these means are energy costly and they would necessarily lead to a lower output of suitably equipped power plants. Moreover, technically, their deployment on a large scale involves unresolved problems, so we cannot rely on carbon capture to reduce future atmospheric concentrations of CO_2.

To ensure the long-term stability of the biosphere, it seems necessary, therefore, to reduce our carbon emissions from 10 Gt to less than 2 Gt per year.

5. From Malthus to COP 21: depletion of energy resources or excessive entropy?

According to the scenarios studied by the IPCC, temperature increase cannot be limited to 2°C unless carbon emissions are compensated for by their capture and storage by 2050. It will be necessary by that time to reduce current emission levels of 10 Gt to a level below 2 Gt of carbon per year. This drastic reduction is the price to be paid in order to recover the balance.

How is it to be achieved within 30 years without causing a brutal disruption of society, and this at a time when the world's population continues to grow quickly? Since the beginning of the oil era in 1950, the world's population has grown from 2.6 billion to 7.5 billion inhabitants. Population has increased by a factor of 3, and carbon emissions by a factor of 6. Has the abundant and cheap energy provided by oil triggered a phenomenon of mass consumption and a rapid increase of population, leaving us wondering today how to bring it to an end? Aren't we experiencing the kind of disastrous collapse scenario foreseen by Malthus and studied by the Club of Rome, but for a reason that has nothing to do with the depletion of natural resources?

The reduction of emissions from 10 Gt to 2 Gt has become necessary due to the climate change triggered by the massive use of fossil fuels since 1950, and not because of their depletion. Although it is not clearly said, the COP 21 Parties recognize that this crisis is different from the crisis commonly called "energy crisis".

As shown, we are witnessing a massive and brutal increase in biosphere entropy due to the massive use of fossil fuels. Beyond temperatures, this increase affects our whole environment. Pollution of air, water and soil, and extreme climate events are only some of the repercussions of this increase.

The scenarios studied by the IPCC, linking carbon emissions and evolution of temperatures, and those studied in *The Limits to Growth* should be considered as a whole. Indeed, in both cases, the rate of exploitation of carbon reserves plays an essential role.

The low estimates of carbon reserves to be exploited "reasonably" amount, according to Turner, to 60.10^{21} joules. This corresponds to a carbon weight of 2000 Gt. At the current rate of emissions of approximately 10 Gt/year, half of these reserves will be exhausted in 100 years, well after 2050. Their exhaustion will have only little impact on their cost of exploitation in the short term, and therefore will not be a reason for collapse. There will be no energy crisis before 2050.

Nevertheless, the COP 21 Agreements require, according to our analysis, the limitation of emissions to 2 Gt/year from 2050 at the latest, in order to prevent temperatures from rising above 2°C. This increase results from the increase of biosphere entropy, which has become the most urgent risk factor.

As Clausius clearly stated, energy and entropy are two fundamental, inseparable notions. We are, indeed, facing two complementary aspects of finitude: material resources, and the lack of empty space to put our waste. This second aspect seems today to be the one requiring a most urgent solution.

Thus, two issues are now arising.

First, we should deal with the question of whether the survival of modern societies really demands maintaining the current level of CO_2 emissions. What would be the optimal reduced level that would cause the least damage — that is, without bringing about a collapse of the economy? We will look at the needs of these societies in Chapter 5.

Secondly, we will study the cost of the implementation of non-polluting energy sources intended to replace the fossil fuels that are to be eliminated. Are there available financial means to proceed to this replacement? An estimate of this cost will be presented in Chapter 6.

Decrease is the first step to take, while doing it progressively to avoid a brutal Malthusian collapse of industrial production and population. The second step is more innovative, but is very costly, as we will see. Necessary investments will no longer be available to develop growth according to the model presently in use.

Is the entropy crisis a programmed crisis? Expected increase in temperatures is the reason why emissions of carbon should be reduced rapidly and drastically, either by saving money or by investing in non-polluting energy sources. But in the end the consequences could well be the same as those mentioned by the Club of Rome: a collapse of industrial production and population. This is the outcome to be avoided.

5

Why do we Release so much Carbon into the Atmosphere?

Based on the work of the IPCC, the COP 21 conclusions urge us to stop adding CO_2 to the atmosphere, a situation that might lead to a climate disaster. Having accepted this conclusion, we have applied the principles of thermodynamics to estimate its cost. Let us recall the stages of this detailed examination.

1. The acceptable limit of carbon emissions

1.1. *The cost of atmospheric CO_2 capture*

CO_2 molecules released into the atmosphere will occupy all the available space. According to Boltzmann, entropy is then increased. Capturing these molecules is generally possible if an external source provides the necessary energy to compensate for this entropy increase, as demonstrated by Clausius.

On this basis, we have calculated the minimal energy required for capturing a CO_2 molecule, or any other polluting molecule, that should be withdrawn from the atmosphere. The result of this calculation is that this energy is equal to the combustion energy of a carbon atom based on a $C+O_2 = CO_2$ reaction.

In other words, the energy produced by the combustion of fossil fuels would be barely sufficient to extract previously released CO_2 molecules from the atmosphere. The ability to capture emitted CO_2

is, therefore, an illusion. Thus, there can be no clean coal. Any CO_2 molecule added to the atmosphere will stay there, at least on our time scale that covers a few centuries. Any extraction attempt would cost a prohibitive price. Therefore, the only way to stop adding CO_2 to the atmosphere is to stop emitting it, or nearly so.

1.2. *The role of photosynthesis in the capture of atmospheric CO_2*

The reason why we do not have to stop completely CO_2 emissions is the important role played by photosynthesis in preserving the balance of the biosphere.

   In principle, solar radiation provides sufficient energy to extract all the CO_2 from the atmosphere that we emit, on a yearly basis. It can, theoretically, extract 1000 Gt of carbon per year, while we emit only 10 Gt. However, the efficacy of photosynthesis in carbon capture is, in fact, very low.

   On the one hand, photosynthesis does more than capture CO_2; it turns it into sugars that feed and enable plants to grow. Given the efficiency of photosynthesis, only 100 Gt of carbon can be captured. This calculation is consistent with the CO_2 cycle measurements illustrated by examples in Chapter 2. In the heavily wooded northern areas (Canada, Russia, and Nordic European countries), the amplitude of this cycle is very clear. During summer, the concentration of atmospheric CO_2 can decrease by more than 10 ppm, which is five times higher than the annual rate of emissions.

1.3. *Stability of the biosphere imposes an emission limit from 1 to 2 Gt carbon per year*

However, as we saw in Chapter 2, these same cycles show that all the CO_2 captured during the summer is to first approximation re-emitted during winter. In order to survive in winter, trees burn the sugars they have accumulated during summer. The annual balance appears to be more or less null.

Nevertheless, a more accurate evaluation of the net balance of photosynthesis can be obtained based on wood production statistics. This production is important, and reaches 2 Gt of carbon per year, or 20% of emissions. But half of the wood produced is burned. The other half is used as construction material. Only 1 Gt of carbon is, therefore, effectively stored, in the medium term at least.

According to the optimistic assumption that all wood produced can be stored at least until the end of the century, and taking into account that the cost of carbon capture by means other than photosynthesis, as we have seen, is prohibitive, we can conclude that the upper limit of carbon emissions compatible with the biosphere's stability is 2 Gt of carbon per year.

This conclusion should be translated in terms of CO_2 emissions per capita and per year. Since there are nearly 8 billion inhabitants on Earth, and the weight of a CO_2 molecule is about four times higher than that of a carbon atom, the authorized 2 Gt (2 billion tons) of carbon per year are equal to 8 billion tons of CO_2 per year.

The level of emissions compatible with the stability of the biosphere is, therefore, *1 ton of CO_2 per inhabitant, per year*. The magnitude of this value should be kept in mind for the rest of our discussion.

2. Distribution of carbon emissions

The global value of carbon emissions hides considerable disparities between the various regions of the world. Before considering how and the methods that should be applied in order to reduce this global value, it is necessary to examine these disparities. We will then try to analyze the elements causing high levels of emissions in the most developed regions.

2.1. *Disparity of carbon emissions: the African continent*

The level of emissions in European countries per capita, per year, reaches 10 tons. In North America, the figure is 20 tons. It might reach even more than 40 tons in Qatar. We shall keep in mind the figure of 10 tons per inhabitant, per year, as an order of magnitude in developed countries. This value, which is ten times higher than the acceptable average value which we have calculated as 1 ton, can serve as a point of comparison with the African continent.

It should be borne in mind that the population of this continent has exceeded one billion — a figure similar to the populations of Western Europe and North America together.

On average, emissions from Africa were 0.32 tons in 2008, 30 times lower than the level of emissions of Western European countries. This value is even much lower than that of 1 ton, which is our acceptable standard. In this sense, the African continent is a virtuous continent. But within this very low average value there are also considerable disparities in the African continent itself.

The level of emissions in a group of countries located in a vast area between North Africa and South Africa, with about 300 million inhabitants, is less than 0.1 tons of CO_2 emitted per capita. This level of emissions is 100 times lower than that of Western countries. For comparison, all these countries emit *200 times less* carbon than the United States of America. How to explain this phenomenal gap? It should be added that, within this group of countries, emissions remained relatively stable from 1960 to 2011.

Yet some African countries reach a level of emissions that exceeds 2 tons.

One country only — South Africa — reaches the emission levels of a typical Western European country (9.3 tons in 2011).

It is also necessary to mention the rapidly increasing emissions in African oil-producing countries (other than Algeria).

African countries with low emissions are characterized by high population growth. The population of Ethiopia which is included in the group of the poorest countries, with an emission level of less than 0.1 tons, has grown from 18 to 100 million inhabitants between 1950 and 2016, i.e. at a rate of 2.5%. The population of Nigeria, where the level of emissions is a little higher, but still very low (0.5 tons), has increased during the same period from 38 million to 200 million inhabitants, with a growth rate of 2.7%. These two countries alone have a total current population comparable to that of the United States, but their rate of growth is much higher. By 2050, their combined populations will have exceeded the population of the European Union.

Table 5.1. Level of emissions in some African countries.

CO_2 emission level in tons per year and per capita	Country
Equal to or less than 0.1	Cameroon, Central African Republic, Chad, Democratic Republic of Congo, Eritrea, Ethiopia, Guinea-Bissau, Mali, Niger, Rwanda, Somalia, Uganda
Greater than 2	Algeria, Egypt, Tunisia
Rapidly increasing	Angola, Equatorial Guinea, Gabon, Libya

2.2. Disparity of carbon emissions: the most polluting countries

The most polluting countries on the planet are at the other end of the spectrum regarding per capita emissions.

Table 5.2 lists the top ten countries in this classification. It is not surprising to see that, in the first place, we find all but one of the major oil and/or coal producing countries, which are in the Persian Gulf, led by Qatar. North America (the United States and Canada) is at the bottom of the list, and Australia is very close. However, a surprising point is that Luxembourg takes fifth place, before the United Arab Emirates, Oman, Saudi Arabia and Bahrain. Luxembourg, which is neither an oil producer nor an industrial power, shows that the disparities of emissions require further study. One might be tempted to assume that it reflects an extremely high standard of living.

Table 5.2. The ten countries with the highest CO_2 emissions in tons per capita, per year (Australia is in eleventh position with 16.5 tons per year (2011, World Bank)).

Qatar	44
Trinidad and Tobago	37.1
Kuwait	28.1
Sultanate of Brunei	24.4
Luxembourg	20.9
United Arab Emirates	20.4
Oman	20.2
Saudi Arabia	18.1
Bahrain	17.9
United States, Canada	16.7

2.3. *Disparities inside Western Europe*

The case of Luxembourg calls for the examination of emission disparities within Western Europe. It is assumed that Western European countries have a similar standard of living and similar climatic conditions. Thus, we would expect a fairly homogeneous level of emissions. But what is really happening there?

Even if we exclude the case of Luxembourg, the level of emissions in Western European countries actually varies by a factor of more than 2 between Holland and Finland, each reaching 10 tons, and Switzerland, where the level of emissions barely reaches 4.6 tons (World Bank, 2011). In some cases, these different levels of emissions clearly reflect government choices. Low emissions in countries such as Switzerland (4.6), France (5.2), and Sweden (5.5) originate from deliberate choices implemented to minimize the use of fossil fuels. Similar efforts are also evident in Denmark, where emissions decreased from 9.8 tons in 1990 to 7.2 tons in 2011. On the other hand, Norway, probably because of its significant oil and gas resources, climbed from 7.4 to 9.2 tons during the same period.

This great disparity within a rather homogeneous zone shows that it is possible to combine a high standard of living with a limited level of emissions. Application means vary from one country to another: massive use of nuclear power (France); wind power (Denmark); hydroelectric power (Switzerland); and measures aimed at improving thermal insulation of buildings. This disparity also reflects the implementation of different industrial policies.

A low level of emissions in one country can result from the import of industrial goods that underwent polluting production elsewhere, and vice versa.

The decision to set aside nuclear electricity has had a negative impact on the evolution of the level of emissions. According to Eurostat data referring to European Union countries, Germany is the only country in Western Europe where the level of emissions has increased from 2010 to 2014. In 2011, eight German nuclear power plants were closed and the proportion of nuclear power fell from 25% to 16%. Coal-fired power plants remain the main source of electricity. The planned final decommissioning of all German nuclear power plants in 2022 might lead to an increase of emissions that would erase reductions achieved by other European Union countries.

Table 5.3 shows CO_2 emission reductions from 2010 to 2014 in ten Western European countries (base 100 in 2010).

Table 5.3. CO_2 emission reductions from 2010 to 2014 (base 100 in 2014).

(Key: FI Finland, DA Denmark, IT Italy, SW Sweden, BE Belgium, UK United Kingdom, NL Holland, FR France, ES Spain, DE Germany.)
Germany is the only country whose emissions have not been reduced during this period, which coincides with the decommissioning of eight nuclear power plants.

FI	DA	IT	SW	BE	UK	NL	FR	ES	DE
79	82	83	85	85	87	88	90	91	101

All Western European countries have significantly reduced their emissions (from 10% to 20%), except Germany, where they have slightly increased. If Germany goes ahead with the closure of the rest of its nuclear plants by 2022, it is doubtful whether its emission reduction commitments as agreed with respect to COP 21 will be fulfilled.

2.4. *Evolution of emissions in major emerging countries*

The major emerging countries are China and India, whose populations exceed one billion inhabitants each (1.383 billion and 1.330 billion respectively). Together, their populations represent more than a third of the planet's inhabitants.

Then comes Brazil, whose population (200 million) is half that of South America, and the Philippines (100 million) as a representative country for South-East Asia.

CO_2 emissions in all these countries increased significantly between 1990 and 2011, much more than emissions in other countries, particularly those in Africa.

Table 5.4 shows the evolution of emissions in these countries from 1990 to 2011 (World Bank).

China's emissions per capita have now reached a level comparable to those found in many European countries. In this sense, China is no longer an emerging country. The situation is still different in India and in Latin American countries. However, the contrast between these countries and Africa, where the average level of emissions remains around 0.3 tons per capita, is striking.

Table 5.4. Evolution of emissions in some major emerging countries from 1990 to 2011 (tons of CO_2 per capita and per year).

	China	India	Brazil	Philippines
1990	2.2	0.8	1.4	1
2011	6.7	1.7	2.2	1.8

2.5. *The various types of fuels used in different countries*

Behind similar emission levels — between, for example, China and an average European country — may lie very different types of consumptions. Indeed, for the same emission levels, a fuel such as coal produces less energy than a fuel such as natural gas. One expresses this difference by saying that coal pollutes more than natural gas. In China, coal burning contributes 75% of emissions, while in Italy it reaches only 10%.

Table 5.5 shows a pollution ranking list for a few typical countries, expressed in fractions of emissions derived from coal combustion.

Table 5.5. Degree of pollution of CO_2 emissions, expressed in % of emissions from coal combustion in 2013.

China	India	Germany	Oceania*	Africa	US Canada	Europe**
73	61	42	39	34	30	22

*Japan, Australia
**Western Europe, not including Germany

Differences are significant. According to this classification, China is the most polluting country, and Western Europe (not including Germany) the least polluting. Germany has a special statute. It is by far the most polluting country of all developed economy countries. The contrast with the rest of Western Europe, which is the least polluting region, is particularly striking.

Figure 5.1 illustrates the evolution of the use of different types of fuels in China. These emissions now account for 25% of world emissions, three-quarters coming from coal.

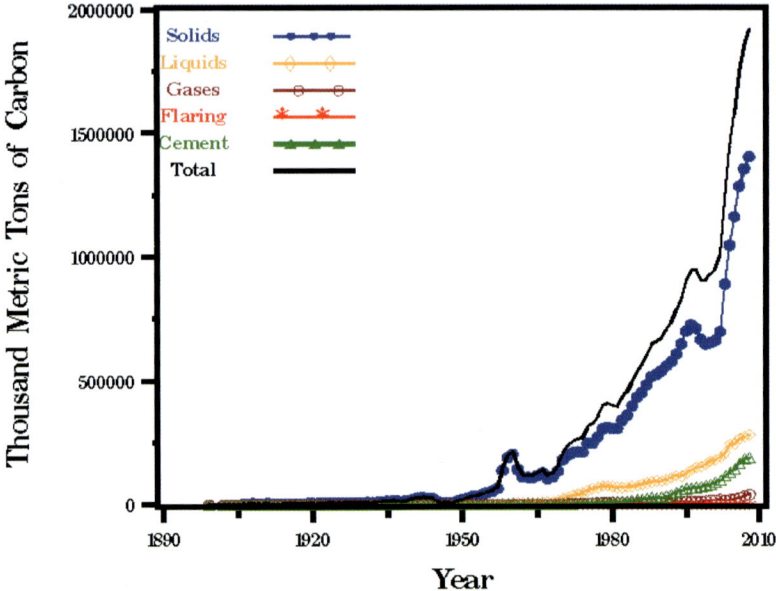

Fig. 5.1. Evolution of emissions in China since the start of the Industrial Era, according to different types of fuels. The rapid rise in emissions since the 1990s is mainly due to the rapid increase of coal production.

2.5. *Should we stop using coal?*

The differences shown in Table 5.5 have in part their origin in natural resources. For instance, China is rich in coal and poor in oil. These differences also reflect a choice between economic development (coal being the least expensive source of energy) and environmental protection. This is the case in Germany, which has significant reserves of lignite that are easily exploitable on the surface. Germany's choice is clear.

Sometimes, global numbers can be misleading. For example, Africa does not use much coal, although, in fact, it is mined with ease in South Africa (78%), which also happens to be the only African country with emissions per capita comparable to those of developed countries.

The use of coal, compared to that of oil or natural gas, offers positive elements, too. While oil revenues are allocated to the capital, those of coal are distributed more efficiently because its exploitation requires a much larger workforce. Coal extraction is harder work than getting oil out of the ground. Coal mining and its use in heavy industry has led — in England, then in Europe — to the development of a skilled workforce, as well as technical mining skills. China has undergone the same process. On the other hand, the exploitation of oil in the major producing countries has had the effect of slowing down social progress.

Today's trend calls for reduced exploitation and the limited use of coal to protect the environment. This concern is clearly justified, but it involves social choices that are not obvious.

Figure 5.2 shows the evolution of emissions in Western Europe (excluding Germany) compared with China.

Figure 5.3 shows, for comparison, the evolution of emissions in Germany. It should be noted that in the three cases mentioned, most emissions originate from the use of coal as an energy source up to the 1950s.

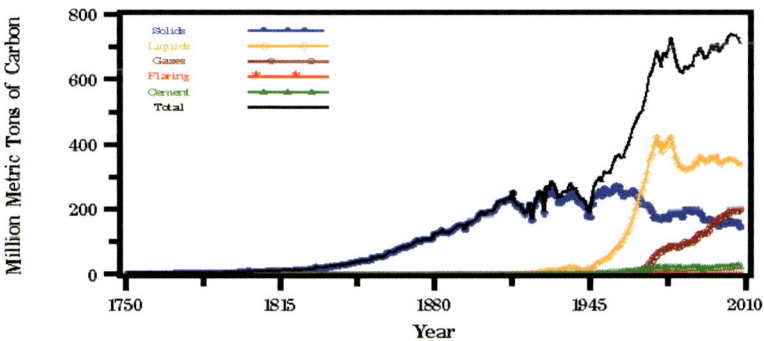

Fig. 5.2. Evolution of emissions (tons of carbon) in Western Europe (excluding Germany) from 1950 to 2013, according to the different types of fuels. Series 2: total emissions; series 3: gas; series 4: oil; series 5: coal. Coal now accounts for only 10% of total emissions. Saturation can be also noted before a decrease in total emissions in recent years.

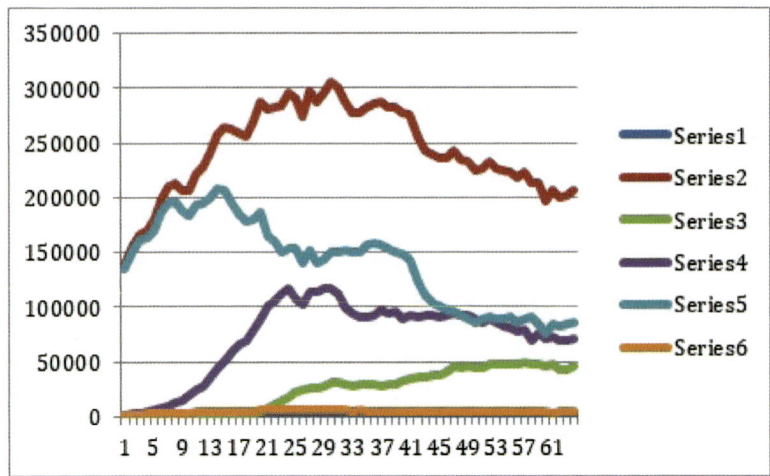

Fig. 5.3. Evolution of emissions (tons of carbon) in Germany. Series 2: total emissions; series 3: gas; series 4: oil; series 5: coal. Unlike other West European countries, in Germany (Fig. 5.2), coal continues to account for a significant share of emissions, almost half, in fact (1950 to 2013).

2.6. *The share of fossil fuels in primary energies*

At this stage, the proportion of fossil fuels used in primary energies should be considered.

Table 5.6 summarizes the evolution of this share from 1973 to 2013. It appears to remain above 80%. There is a slight decrease from 1973 to 2010, mainly because of the growing part played by nuclear power, but it has stabilized since. Apart from biomass (wood, mainly), the contribution of other renewable energies remains marginal, at around 1%.

Thus, in order to ensure the stability of the biosphere, carbon emissions need to fall from 8 Gt to 1 Gt per year (unless produced wood is no longer used as fuel, but is fully stored, which is unlikely to happen). In this case, available primary energy would amount to only 30% of what it is today, while all other sources remain constant.

How could we make up for this dramatic decrease? An increase in nuclear power is uncertain, especially in these proportions. Is the evolution of our modern societies compatible with such a reduction in energy consumption? This will be the core question studied in the last chapter of this book.

Table 5.6. Global production of primary energy.

In Mtep	1973	%	2010	%	2013	%
Coal + lignite	1474	24.5	3476	27.3	3958	28.9
Petroleum	2938	46.2	4107	32.4	4216	31.1
Natural gas	991	16.0	2728	21.4	2909	21.4
Nuclear	53	0.9	719	5.7	646	4.8
Hydropower	110	1.8	296	2.3	326	2.4
Biomass + waste	640	10.5	1278	10.0	1376	10.2
Other*	6	0.1	114	0.9	164	1.2
TOTAL	6213	100	12,717	100	13,594	100

*Others: Geothermal, Solar, Wind, Recovered Heat, etc.

2.7. The impact of emissions reduction on food production

Traditional agriculture uses solar radiation (photosynthesis) as its main source of energy, apart from the physical work of the farmers themselves and their draught-animals. Therefore, its impact on total fossil fuel consumption is negligible.

On the other hand, modern agriculture, as practiced in developed countries, uses fossil fuel-based energy sources. This energy supply — mainly in the form of fuel used for seed and crop machines, drying and storage processes, transport, and production of fertilizers and pesticides — has led to an enhanced yield per hectare. The consumption of fossil fuels in agricultural production is no longer negligible.

Thus, the question is: what would be the impact on agricultural production of a massive reduction in the use of fossil fuels, aimed at restoring the balance of the biosphere? Is the level of emissions compatible with this balance, 1 GtC per year, able to maintain a level of production sufficient to ensure the food supply?

A person consumes approximately 2500 calories of food, or 10,000 kilojoules (kJ), per day. For a population of 8 billion, it represents a consumption of $2.5,10^{19}$ joules, or 5% of global energy consumption. This is significant.

The use of fossil fuels, aiming to increase agricultural yields in developed countries, is of a similar amount. Additional energy used in the processing and packaging of products should also be taken into account. Each country and each region has its own characteristics, and numbers are not clear. But this energy is considered to be of the same order as the consumption of energy used directly in agricultural production.

Overall, it can therefore be estimated that 10% of total CO_2 emissions are related to the production and consumption of agricultural products, meaning around 1 GtC per year. This level is compatible with the balance of the biosphere.

The answer to this question is that keeping the level of agricultural production at its current level requires the use of all fossil fuels compatible with the biosphere balance.

In a balanced biosphere, industry and transport (except those involved in agricultural production) and residential uses (lighting, heating and cooling) will no longer be able to use fossil fuels.

3. Entropic waste and the complexity of modern societies

This conclusion calls for a review of the way our modern societies function, since the balance of the biosphere requires a return to a lifestyle similar to that of undeveloped societies, where agricultural production represents the core of human activity. If this is to be avoided, we must develop massive alternative energy sources for industrial, transportation and residential uses. This would require considerable investment. It will, therefore, be necessary to reduce present consumption, whilst keeping that which is essential for maintaining the progress made since the start of the industrial era.

We must, therefore, return to the question raised previously: how is it that, in our developed societies, consumption is 100 times higher than in the poorest countries? Is this huge consumption really essential, and what is it exactly?

It should be noted that this consumption is not mainly due to the use in private sphere. Contrary to the prevalent argument, the savings we can realize by reducing consumption, controlled by us in our daily lives, are only a small part of global consumption per capita.

3.1. *Energy consumption in the family*

Awareness campaigns to encourage or even force us to reduce our energy consumption generally focus on, for example, domestic thermal insulation efficiency, lighting, and home appliances. Accurate standards are imposed to eliminate incandescent lamp lighting and improve the energy performance of household appliances such as refrigerators, ovens, and toasters.

But what can we really expect from these measures? Are they compatible with the actual issue at hand? Can these *individual* measures help to sharply reduce CO_2 emissions?

The answer is rather simple. We should make a list of the energy expenses that we can control within our homes. An estimate is given in Table 5.7. In order to be able to compare the different forms of energy used to cover these needs — sources that can be provided in all kinds of forms (electricity, fuel, domestic gas, and petrol) — all the needs are expressed in kWh. Food needs are calculated on the basis of 2500 calories per day, per person.

The figures, for information purposes only, apply to a family living in Europe. Obviously, needs vary from one family to another according to their standard of living (the size of their refrigerator, for example), the amount and quality of their home's thermal insulation, the number of cars they own, and whether they use gasoline or diesel, and so on. What we must retain is the order of magnitude of an average family's consumption: 100 kWh per day, 10 times more than the food requirements which are the most essential and incompressible.

Table 5.7. Daily energy needs for a family of 4, in kWh.

Food needs	
Based on a consumption of 120 W per person exercising minimal physical activity	10
Home appliances	
Fridge, washing machines, television, computers, lighting	10
Hot water	
Sanitary needs	10
Heating/Air Conditioning	
For a well-insulated residence of 160 m^2, consuming approximately 1000 liters of fuel or equivalent per year	30
Transport	
For a vehicle traveling 20,000 km per year, consuming an average of eight liters of gasoline per 100 km. One liter of gasoline contains about the equivalent of 10 kWh in the form of chemical energy.	40
TOTAL	**100**

We note that the individual family's consumption, or 25 kWh, is equal to 1 kW of power, which is five times lower than the overall consumption per capita in the context of a developed European society, and ten times lower than in North America.

Thus, consumption per capita corresponds to a power of 100 W for food requirements, a power of 1000 W for an individual's consumption in a family frame work, and 5000 to 10,000 W at the society level in a developed country.

In terms of CO_2 emissions, 100 W corresponds to the emissions per capita in the poorest countries (less than 0.1 tons per year), 1000 W to the emissions per capita of developing countries, and 10,000 W represents the emissions in the highest-consuming countries. The unsustainable level of emissions is not due to consumption within the family, but consumption within the framework of society. In our societies, 80 to 90% of emissions are due to consumption at societal level, which we cannot control as individuals.

It is certainly virtuous to reduce the emissions that we control directly, but, in order to achieve the drastic reduction in emissions necessary to regain the biosphere's balance, societal measures should be taken. This point will be discussed in the next chapter.

3.2. *The energy cost of food*

In any case, the role of nutrition will always remain essential in any society. The fact that it represents only about 10% of our direct energy consumption in the family is misleading. The cost of 1 kWh of food is generally much higher than the cost of 1 kWh of electricity, and this is for two reasons. First, the production of staple foods — cereals, rice, etc. — requires an external energy supply (fertilizers, pesticides) that completes the photosynthesis process. Secondly, and this is the important point, most of the foods we eat have undergone changes that very significantly reduce energy efficiency.

Table 5.8 illustrates this situation. The energy cost of foods that have undergone minimal processing — for instance, rice and spaghetti — is slightly higher than the cost of electricity (0.1 to 0.2 euro/kWh depending on the country), but remains of the same order. However, the energy cost of meat is much higher.

These wide price differences in terms of consumption expressed in Euros/kWh reflect a physical reality. A calorie of rice has the same nutritional value as a calorie of beef, but its energy cost is 50 times lower. In other words, its impact on the environment is, at first approximation, 50 times smaller. A society that consumes its majority of calories from beef will pollute the environment 50 times more than a society where most of the calories consumed originate from rice or pasta.

This figure is similar to the difference in magnitude between per capita emissions in developed societies and the poorest societies.

Table 5.8. Energy cost of some foods in Euros/kWh.

Rice, spaghetti	0.3
Bread	1.1
Potatoes	1
Tomatoes	12
Chicken	8
Beef	17

3.3. *The budgetary impact of food on a family of four persons*

The share of food involved in energy consumption is only around 10% (Table 5.7). But in terms of budget, this same share becomes dominant. Our energy expenditure on food accounts for about half of our total energy expenditure, as shown in Table 5.9. It is largely due to the low energy yield of our mode of nutrition.

Table 5.9. Daily expenses for different types of energy consumption in Euros for a family of four persons.

Food	15
Household appliances	1
Hot water	1
Heating	4
Transport	6
Total	27

In this table, the share of food has been calculated on the basis of an energy cost of 1.5 Euro/kWh. This cost is on an average between the price of rice (0.3 Euro/kWh) and the price of meat

(17 Euros/kWh). Thus, these numbers relate to a rather modest family, meat representing only a small part of the calories consumed.

The weight of food in the overall energy balance, or in the weight of emissions, is important. A drastic reduction in emissions will probably require a change in nutrition habits among developed societies.

6

Means for a Recovery of Balance

As stated by Clausius, energy and entropy are two distinct, but complementary and intrinsic, notions. As far as life is concerned, a study of the combination of the two is particularly essential in order to understand the limits of its development. These limits, which we are facing today, are as much about energy as about entropy. In order to emphasize how essential it is to deal simultaneously with the energy aspect and the entropy aspect, we propose, in the spirit of Clausius, that the International Energy Agency (the IEA) changes its name to the *International Energy and Entropy Agency*, to become the *IAEA*.

1.   Limits to growth: energy and entropy

Life requires an ongoing supply of energy to function, to grow and to reproduce. Without this supply, life would stop. We are aware of this, which is why the prospects of a lack of energy in the future causes so much anxiety. The works of Malthus and the Club of Rome are the relevant reasonable expressions of this issue.

On the other hand, the very development of life pollutes its environment. This is the second limitation that it faces. The functions of living generate increased entropy. Increasing entropy in our environment appears in various forms of pollution, such as greenhouse gases, air pollution and water and soil pollution.

Greenhouse gas emissions attract most of the attention today, but there are other forms of pollution that have become a more immediate threat to our daily life. According to the World Health Organization (WHO), these forms of pollution cause millions of premature deaths every year.

Growth can be limited by a lack of energy as well as by an excess of entropy. These are two different types of finitude. In the first case, growth turns into a recession due to lack of resources, as foreseen by Malthus and the Club of Rome. In the second case, it also turns into a recession, but for quite another reason: economic activity is slowed down by the accumulation of waste that disrupts the regular functioning of the biosphere.

Let us take a concrete example: traffic can be stopped either because of a lack of fuel, or because pollution has reached a stage deemed a danger to health. The second case involves a decision to be taken by the authorities. It might be controversial and difficult to take, as it causes immediate loss of economic activity, while health hazards and measures taken to deal with them are less obvious.

1.1. *The IPCC and COP 21 approach*

IPCC experts have modeled exchanges between the atmosphere, emerged lands and oceans, and managed to link the evolution in CO_2 content with the rise in temperatures. They focused on general warming, which is an entropic effect since it comes from the accumulation of greenhouse gases.

Counting from the start of the industrial era, this increase has been slow, about 1°C over a century. Modeling predicts an increase of the same order by the end of the 21st century. It could even double or triple, depending on the evolution of emissions. This was considered sufficiently dangerous by COP 21 for them to recommend — or even eventually impose — a sharp reduction in the use of fossil fuels.

But the argument they have used can be questioned. As can be seen from past surveys, temperature increase is not as steady as the evolution of CO_2 content. It has undergone significant fluctuations, a few tenths of a degree. With the focus on a forecast of rising temperatures, the IPCC and COP 21 may be exposed to criticism. It might very well be that during a few years, temperatures will remain as they are, or even decrease slightly, as has happened in the past.

As presented in Fig. 2.1, temperatures increased by 0.6°C between 1910 and 1945, but then remained stable until 1970. Over a shorter period, from 1945 to 1950, they decreased by 0.3°C. If this were to happen again, industrial lobbies — naturally climate-skeptic — would not hesitate to use this fluctuation to question the warming itself.

While going in the right direction by recognizing the importance of a phenomenon that is not related to the prospects of a lack of energy resources, the IPCC experts' approach did not take into account that rising temperatures are only one of the aspects of the entropy crisis, and not necessarily the most dangerous in the short run. Greenhouse gas effect claims far fewer victims today than air pollution by micro particles, another effect of entropy. Extreme weather events, which are not forecast, can do much more damage than a slow increase in temperature. They are also a natural consequence of entropy increase.

1.2. *Thermodynamic approach*

Our approach is fundamentally different from that of the IPCC and COP 21. We have evaluated the maximum level of emissions compatible with maintaining the global balance of the biosphere, based on a law of thermodynamics according to which increases of entropy due to these emissions may be offset by an external energy supply.

We have then calculated the excess of entropy in the biosphere resulting from the increase of CO_2 concentrations in the atmosphere, while putting aside all other forms of pollution — not because they are less important, but because they are more difficult to quantify. Our estimate of emissions compatible with maintaining the stability of the biosphere is therefore an upper limit, a rather optimistic one, of these emissions.

We proceeded then to the calculation of the energy input necessary to ensure the stability of the biosphere, and we compared it to the energy supplied from the sun through photosynthesis. We have reached the conclusion that this contribution of solar energy is largely insufficient to compensate for the increase of entropy at the current level of CO_2 emissions. While these emissions are of the order of 10 Gt of carbon per year, the energy received from the sun by photosynthesis can only compensate for up to 2 Gt of carbon emitted per year, and perhaps even only 1 Gt of carbon, if half of the harvested wood is used for heating or cooking, as is presently the case.

Therefore, in order to preserve the balance of the biosphere, it is, in fact, necessary to stop using all fossil fuels for industry, transport and building purposes. The use of fossil fuels should be limited to agriculture, as they are essential for maintaining food production at its current level. This use corresponds to emissions of the order of 1 Gt of carbon per year, which is precisely the authorized limit.

This conclusion implies drastic changes in the behavior of developed societies, which emit most of the greenhouse gases. We will now examine them. It should be noted, however, that emissions of the order of 1 Gt carbon a year are comparable to the level of emissions in developing countries, and are even ten times higher than those found in the poorest countries, such as Mali or Eritrea.

2. Origins of high consumption at the society level

We have seen that consumption directly under our control represents no more than 10% of consumption on a societal level. Even if we were to reduce this individual consumption by a factor of 2 or more, it would not lead to a significant change. Individual frugality being intended to limit our consumption of electricity, heating fuels and gasoline would not be a solution.

It is, therefore, essential to try to understand why we, the inhabitants of developed countries, emit ten times more at a societal level than we do as individuals. What is the origin of this amazing gap?

It is up to sociologists, rather than physicists, to answer this question. However, we can make two obvious observations.

A first one is that our individual "energy budget," as discussed in the previous chapter, does not take into account emissions that, although not under our direct control, still depend on our behavior and habits of consumption. We will refer to them as emissions that are indirectly related to our mode of consumption.

The second one is that residents of developed countries enjoy a protective and stimulating environment that is non-existent in poor societies because they do not have the means to develop it for their communities. This environment has a cost in terms of emissions at a societal level.

2.1. *Emissions indirectly related to our mode of food consumption*

A first example of emissions that are indirectly related to our mode of consumption concerns food. Although this issue has already been pointed out in the previous chapter, a few more remarks are worthwhile.

While our individual "food energy budget" represents no more than 10% of the total budget when we measure it in terms of calories or kWh, as shown in Table 5.7, it accounts for half of it when expressed in monetary terms as seen in Table 5.9. This considerable difference reflects the low energy yield of food, i.e. the energy contained in the food is low compared to the energy that was consumed to produce, transport, store and distribute this same food.

Our actual food energy consumption is much higher than the number of calories we buy.

Here the consumer's habits and the means at his disposal have a role. As seen in Table 5.8, the price of a calorie can vary significantly. This reflects the great variability of the food's energy yield. It is highest for staple foods such as rice and pasta, which provide the cheapest calories. It is low for meat, and even lower for exotic foods like some vegetables or fruits that cross the oceans before being put on the market.

It is, therefore, very difficult to evaluate the real individual food energy budget. An average value will not make sense, as this budget can easily vary by more than an order of magnitude within a developed country.

For instance, in many developed countries, there are social programs intended to enable members of the poorest communities to eat their fill. In the United States, for example, the Food Stamp Program, currently known as the Supplemental Nutrition Assistance Program (SNAP), enables those with incomes below a certain limit to purchase food for $4 per day, per person. This $4 budget can certainly provide the number of necessary calories, but not the famous so-called essential five fruits and vegetables, however.

It is clear that, in well-to-do families, the actual food budget is much higher than $4 per day, per person. This budget implies high indirect emissions that we are not aware of at the individual level, but which are accounted for at the level of overall emissions.

2.2. *Indirect emissions related to the purchase of consumer goods*

The household energy budget shown in Table 5.7 does not take into account the energy impact involved with the purchase of a wide variety of consumer goods. Any manufactured product bears an energy cost. Its production, transport and distribution result in CO_2 emissions and different types of pollution. The more we consume, the more we emit, and the more we pollute.

These emissions are not under our direct control, but we are, nevertheless, responsible for our consumption pattern. Unlike direct energy costs such as heating or transportation, we are unable to quantify this cost. We only know the prices we pay, and we have no way of understanding the indirect costs of our purchases in terms of emissions and pollution.

For example, many consumer goods purchased in Europe or the United States are imported from Asia because they are cheaper for the consumer compared to locally manufactured goods. In addition to the lower wages paid in Asia, another reason for this is less demanding production standards. These imports lead to an increase in not only local pollution, but also in global emissions.

The problem is that an adjustment to our mode of consumption, be it food or consumer goods, would immediately affect the functioning of the economy. The reestablishment of customs barriers, which is on the agenda today, would lead to an increase in prices that would certainly cause a negative reaction.

2.3. *The impact of the digital revolution*

The digital revolution is an essential element of modern life. However, it has a negative impact regarding our emissions, an impact of which we are unaware, and is completely out of our control.

It is not just about the power consumption of our personal computers; we pay no attention to that, and if we want to know, it is easy to calculate.

The main issue here is the use we make of the Internet. All the mail, documents, photos and videos that we share pass through servers that consume a lot of power, to such a degree that their cooling systems are a fundamental part of their operational costs. Every item stored in the Cloud implies an energy cost, and causes CO_2 emissions at societal level.

Emissions that are indirectly related to our consumption, and are not under our direct control, are extremely variable from one level of society to another, and from one country to another. We have seen a few examples. These emissions are likely to be more significant than the emissions that are directly under our control. There is undoubtedly room for reflection on this subject, and finding a way to reduce emissions that would not undermine the foundations of modern society. Food habits, purchase of consumer goods, and our use of digital technology could all be changed without affecting the essential positive aspects of modern life.

3. Developed societies in protective and steering environments

High societal emissions in developed countries also originate from the outstanding services provided to their residents. These services are not found at all in less developed countries, and only partially in developing countries.

Such services include hospitals, schools and universities, roads, an established administration, police, public transport, infrastructures that provide access to drinking water and electricity, sewage management and many others. Although we take them for granted, these services are not obvious. They represent considerable investment and maintenance costs that result in CO_2 emissions and various types of pollution. Unlike emissions related to our (bad) habits as consumers, we consider these services to be essential to the proper functioning of our societies. These services are not available in poor countries.

3.1. *Structures for education and health*

Education and health are among the services we consider essential. These services are based on highly complex structures.

Children's education starts at a very young age, in kindergarten and often even before, and can continue well beyond the age of 20 for those going on to higher education. The education system has become very competitive, with the most talented children promised the best future. Although imperfect, this system generally encourages heavy personal investment. It offers the individual who is a part of it the feeling that an open world awaits, enabling her/him to realize her/his potential.

There is no finitude here. Education at its pinnacle leads to research, which knows no limits. This type of growth through study concerns both the individual and society. It requires the establishment of a very complex structure that involves an army of multi-tasking teachers at very different levels, from kindergarten to Nobel Prize-level laboratories.

In contrast, poor countries can, at best, give their children a basic education for only a few years. Children born in Mali have little chance of participating in the journey of opportunity open to those who were lucky enough to be born in a developed country. Their only chance is to leave their homeland.

Health system structures in developed countries are not less complex. Whether dealing with nurses or specialists able to handle the most complicated cases, the system provides patients with a range of care that is inaccessible for people in poor countries. They may not even know that this type of care exists.

3.2. *Inequalities in developed countries*

Even in developed societies, access to health and education systems is not the same for everyone. Inequalities are just as glaring as those between developed and poor countries. They may be even less acceptable, because they are more visible.

Despite the existence of the necessary structures, a substantial fraction of the population of developed countries has access neither to upper (or even elementary) education, nor to the best-performing healthcare (or even elementary healthcare, as in the United States). The fear to be found among those excluded is a great source of anxiety for many.

Is this an inevitable polarization in developed societies? Is it a sign that growth can no longer apply to all, but is accessible only to a few? Today, it has become a general feeling. As we have seen, the mode of growth that developed countries experienced between 1950 and 1980, known as the Thirty Glorious Years, cannot actually continue. To take a recent example: the export of household waste from Italy to Morocco is not a solution for the future. Moreover, this mode of growth cannot extend to developing countries.

However, the progress we cherish most — that is to say, in education and health — does not generate the most waste.

There is no doubt that it is up to developed countries to set an example for societies to evolve in a way that excludes no one.

4. A necessary transition

The need for a transition is, therefore, prominent, and the recent evolution of CO_2 emissions should be dealt with first.

4.1. *Emissions from 1950 up to now*

We have set out that the level of carbon emissions compatible with maintaining the biosphere balance is 2 GtC per year at most. This threshold was crossed around 1950, at the dawn of the famous Thirty Glorious Years.

Let us go back to the evolution of emissions since that date.

The use of coal increased rapidly until 1920, and then stagnated until 1950. It was the most exploited fuel up to that point. Then there was a take-off of global emissions that coincided with an exponential increase in the massive use of oil. This phase ended around 1980 — precisely the Thirty Glorious Years. Within 30 years, oil emissions passed from almost nothing to 2.5 GtC.

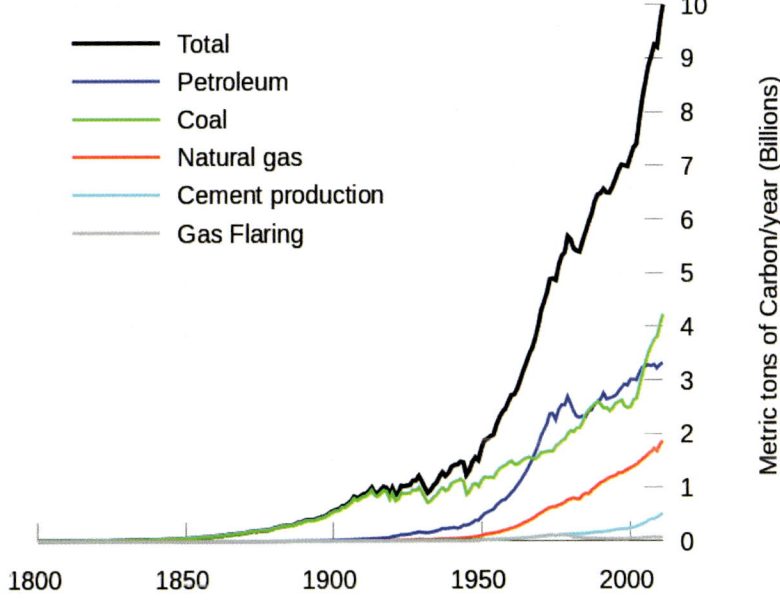

Fig. 6.1. CO_2 emissions since the start of the Industrial Era are expressed in gigatons of carbon. The threshold beyond which the balance of the biosphere is broken, which is 1 to 2 Gt, was crossed around 1950, due to the very rapid increase of oil consumption.

Oil emissions then rose more slowly, from 2.5 to 3.2 GtC in the next 30 years. There was also only a moderate increase of global emissions from 1980 to 2000.

This was where a second phase began, marked by a rapid increase in emissions due to coal consumption, which was making a noticeable comeback. From the year 2000, within fifteen years, emissions related to coal increased by around 2 GtC, at a rate comparable to that seen at the start of the Thirty Glorious Years. This happened mainly in China.

4.2. *Projected future emissions*

Indeed, an examination of the evolution of emissions since 1950 shows that it is not marked by a steady increase, but proceeds in stages. Table 6.1 summarizes this evolution.

The first group of regions that, around 1980, reached a level of 10 tons of carbon per capita and per year, includes North America, Europe and Japan, whose populations together account for about 1 billion people. In 1960, almost all emissions came from this group. In 1980, a major part of the global emissions of 5 GtC still came from these countries. Their emissions have actually stabilized in the meantime, and have even begun to decrease.

China, with 1 billion inhabitants, then joined this first group. Emissions increased gradually at first, reaching a level of 1.5 tons per inhabitant around 1980. After 2000, they increased very quickly and, today, equal to those of the first group. China's emissions have not yet stabilized.

Then comes India, with a third-of-a-billion inhabitants. Their emissions today reach 2 tons, which was China's situation in 1980, and it is rapidly increasing. We can, therefore, expect India's emissions to make a greater leap, comparable to that of China, reaching the same level as the first two groups in 20 years. It will be this third-of-a-billion people who will reach the same level of emissions as developed countries. This increase will probably result from the use of coal, which is by far the cheapest source of energy. If this scenario occurs, following a plateau in emissions during the next decade due to lower emissions in the first group and then

China, a new phase of rapid increase could occur and reach 15 GtC by 2030.

Table 6.1 provides a more detailed picture of emission trends with a projection toward 2030. Combined emissions from the United States, Europe (including Russia) and Japan have not changed much since 1960.

Table 6.1. Evolution of the compared emissions of three major regions, which each include about 1 billion inhabitants.

	1960	1980	2000	2015	2030
USA + EU + RU + JP	3	3.3	3.5	3.3	3
+ China	3	3.5	4.5	6	7
+ India	3	3.6	4.8	6.9	9
TOTAL	3	5.5	7	10	15

In 1980, emissions from China became noticeable, and the effect of the rapid increase in oil production was felt globally as compared to 1960.

In 2000, emissions from China and India (1.3 Gt) accounted for about one-third of those in the first group.

In 2015, they exceeded them (3.6 Gt). Combined emissions of the first group, China and India, accounted for 70% of global emissions, while their populations represented about 50% of the total population.

The rapid increase coming from India echoes the increase that occurred in China at the end of the last century. In the future, emissions from India might, in 2030, be equivalent to those of China today. Three billion people would emit together 9 GtC per year, while Earth's other 5 billion inhabitants would account for 6 GtC.

In the long run, emissions from other major groups such as South America, South-East Asia and Africa could also become important, in successive waves.

Indeed, humanity is the same everywhere. There is no shortage of coal. What has been achieved in China is now happening in

India. Why not elsewhere later on? If each billion people emit 3 GtC per year when all countries have developed, we could reach around 2050 a level of emissions higher than 20 GtC.

This is a disaster scenario regarding the balance of the biosphere. Can it still be avoided, and if so, how? And at what price?

5.   The cost of transition

A transition to a largely carbon-free society is, therefore, essential. This point of view is widely shared today.

We can conceive that measures intended to reduce emissions by improving energy efficiency in all areas can, alone, bring down levels of emissions from 20 GtC to 10 GtC. But such measures cannot in any way reduce emissions to 2 GtC.

5.1.   *Provision of non-polluting energy*

Thus, non-polluting sources that can provide the same energy as obtained with burning fossil fuels emitting 8 GtC per year should be developed.

Total energy consumption today is 100,000 TWh of primary energy. (One terawatt hour — TWh — equals 1000 gigawatt hours, the energy supplied by one thousand nuclear power plants for one hour.) Fossil fuels provide 80% of this energy. Taking into account the level of emissions compatible with the balance of the biosphere, it is necessary to find 64,000 TWh of primary, non-polluting energy — moreover, energy involving an exploitation method that does not generate CO_2.

Nevertheless, non-polluting energy does not mean renewable energy. Wood is renewable, but is still a polluting energy, like fossil fuels, and probably even more polluting because the energy efficiency of wood combustion is lower than that of coal, oil or natural gas. Its combustion releases more CO_2, than they do, not to mention large quantities of micro particles.

Likewise, biofuels and any form of biomass are not sources of non-polluting energy. Hydroelectric, wind, solar and nuclear energy are the only ones that are non-polluting. They alone will have to provide the required 64,000 TWh. Even then, part of the energy released by these sources must first compensate for the energy necessary for their construction and maintenance.

In the case of nuclear energy, it will also be necessary to deduct the cost of waste storage and the cost of dismantling reactors at the end of their lives.

5.2. *The cost of the variability of wind and solar energy. The German model*

Due to their variability, wind and solar power alone cannot provide enough energy. They must be backed by conventional power plants to cope with the demand for electricity in the absence of wind and sunshine.

This is the actual situation in Germany, a country that has invested heavily in renewable energies. Though they will represent a prominent fraction of capacity, as planned, each renewable MW will have to be doubled by 1 MW available at any time so as to ensure the supply of electricity in all circumstances.

This German model is not sustainable. Wind and solar installations provide power only 20 to 30% of the time. Other sources of energy will, therefore, have to cover 70 to 80% of the demand for electric power, which will be greater than today, because transport will also have to be carbon-free. The German model rejects nuclear energy, and hydroelectric power can hardly be developed more than it is already. Therefore, it will be necessary to build new thermal power stations. In total, fossil fuel consumption will only be reduced by 20 to 30%. Moreover, thermal power plants might use the cheapest fuel — that is, coal — which is a matter of concern.

The cost of renewable MW is not the determining factor here. It is rather the cost of its variability that matters.

5.3. *The need for storage of electrical energy*

A truly carbon-free economy is, therefore, only possible if renewable and variable energies are stored in such a way as to be available at all times. Only if their storage is included can the cost of renewable energies be realistically estimated.

Different solutions seem possible, and are presently being studied. One of them relies on using part of the electricity produced to make hydrogen and store it either as gas, or liquid form, or in the form of a compound such as palladium hydride. Hydrogen production, storage, transportation and distribution would require the construction of a new infrastructure with a cost yet to be established.

The cost of electricity storage in accumulators is easier to evaluate. It is not necessarily the best solution in the long run, but such accumulators are now commercially available for storage capacities that correspond to a day of consumption at the household level, meaning a few dozen kWh. A pilot scheme on a small scale is possible.

5.4. *Cost of a global photovoltaic system including 24-hour storage*

This example demonstrates a system integrating photovoltaic (PV) power generation and its storage during a period of one day.

The data is as follows:

COST OF THE PHOTOVOLTAIC SOLUTION, STORAGE INCLUDED, ON A GLOBAL SCALE
Cost of the photovoltaic installation: 1 euro per installed watt.
Average sunshine: 5 hours a day.
Cost of storage: 0.7 euro per watt hour.
Integrated cost of production and storage: 1 euro per watt hour.

Energy required to store globally for one day: 60,000 TWh/365.
Number of inhabitants: 7 billion.
Necessary investment per inhabitant: 25,000 Euros.
Annual cost assuming depreciation of installation over 10 years: 2500 Euros per inhabitant, per year.

This cost is out of reach for developing countries.
Another estimate can be made, to take into account that only countries of the first group will have the means to make this investment.

COST OF PHOTOVOLTAIC SOLUTION, STORAGE INCLUDED, FOR DEVELOPED COUNTRIES
Cost of the photovoltaic installation: 1 euro per watt installed.
Average sunshine: 5 hours a day.
Cost of storage: 0.7 euro per watt hour.
Integrated cost of production and storage: 1 euro per watt hour.
Energy required for storage for one day: 20,000 TWh/365.
Number of inhabitants: 1 billion.
Required investment for a carbon-free process in the first group of 1 billion inhabitants: 60,000 Euros per inhabitant.
Cost per inhabitant, per year: 6000 Euros.

A similar result was obtained by Legoupil (Energy Transitions, Paris-Tech Alumni, 2016), for five-day storage using conventional lead batteries.

The GDP per capita in the countries of the first group amounts to several tens of thousands of Euros per year. These countries can afford to cover the cost of their carbon-free process, provided they allocate 20% of their GDP to it.

This is not the case for the second billion inhabitants. GDP in China is less than 10,000 Euros per year. Thus, this carbon-free model is out of reach for China.

De-carbonation of the third billion people (India) is even less feasible for India — not to mention Africa. This is unlikely to change even in the long term. In other words, carbon-free transition must be gradual, starting with the richest countries.

5.5. *Estimated cost of one carbon-free kWh*

The cost of this solution can also be represented by the price per kWh to be paid by the consumer. The relationship between the cost of production and the price paid by the consumer is rather complex. It generally involves a large number of factors in addition to fuel cost, and investment in maintenance, transportation and distribution. In addition, it involves political decisions.

In the case of the PV storage solution, the cost of the fuel is zero from a local point of view, meaning the place of production is close to the place of consumption. In this case, we can disregard transport and distribution costs. The table below does not take into account cost of maintenance, having also retained a 10-year depreciation, whereas PV panels are operational for about 30 years.

This table gives estimates of the price of one carbon-free kWh according to a chart that considers the latitude and the mode of production, either individually (cost of PV panels 2 Euros/Watt peak, storage cost 0.7 Euro/Watt hour) or collectively (cost of PV panels 1 Euro/Watt peak, storage cost 0.2 Euro/Watt hour).

Table 6.2. Price per kWh PV + storage 24 hours, in Euros.

PRODUCTION	LOW LATITUDES Less than 30°	AVERAGE LATITUDES (e.g. PARIS)
INDIVIDUAL	0.30	0.47
COLLECTIVE	0.16	0.32

In low latitude countries, in collective production mode (about 100 dwellings) the price per kWh is attractive. On the other hand, in high latitude countries and in individual production mode (PV panels on the roof), it is prohibitive.

6. Conclusions

Regaining the balance of the biosphere requires a massive transition to carbon-free production mode. It will have to employ renewable, clean energies.

It is imperative that these energies be supplemented by storage facilities to ensure the regular supply of electricity despite intermittent source.

We have evaluated the cost of a solution combining a renewable energy source — electrical power produced by photovoltaic panels — and storage. These are rough assessments. What one must pay attention to is the order of magnitude. The true cost of renewable, non-polluting energy made available at any time by storage amounts to tens of thousands of euros per year, per household. It should be emphasized, however, that this sum covers all energy needs at a societal level, including the energy required for transportation, heating and industry.

Most of the cost is due to storage, rather than to generation of electricity.

Without storage, renewable energies will not fulfill expectations. On the contrary, the constraints they impose on the network when used without storage can cause disruption, with supply interruptions and high costs, which will be unevenly borne by the various economic factors. The German example, according to which this weight is borne by households, and not by industry, gives an idea of the distortion that this massive implementation of a renewable energy policy can cause when not applied in conjunction with storage.

Rich countries like Germany might be able to become carbon-free, by adding storage to their renewables. In fact, they should attempt this venture quickly.

An alternative solution to massive electricity storage would be a revival of nuclear energy intended to provide about half of the needs. This might be possible with fast breeder reactors. However, the general opposition to nuclear energy and the associated technical difficulties are such that we cannot think about this solution for decades to come.

BIBLIOGRAPHY

An essay on the Principle of Population, Thomas Malthus. F. Johnson, London, 1798.

The limits to growth, D.H. Meadows, D.L. Meadows, J. Randers, W.W. Behrens III., (Universe Books, New York, 1972.)

A comparison of limits to growth with thirty years of reality, Graham Turner. CSRIO Working Papers Series, 2008.

Qu'y avait il donc dans le fameux "Rapport du Club de Rome", J.M. Jancovici.
Manicore.com, 2009.

Réfl exions sur la Puissance du Feu et sur les Machines propres à développer cette puissance, Sadi Carnot, 1824. Re-impression Editions Gabay, 1990.

The mechanical Theory of Heat with its applications to the Steam Engine and to physical properties of bodies, Rudolf Clausius. John van Voorst, London, 1865.

The Entropy Crisis, Guy Deutscher. World Scientific, 2008.

Consommation d'Energie et Rejet d'Entropie dans la Biosphère, Guy Deutscher.
Reflets de la Physique **40**, 31, 2014.

Kicking the oil addiction: facts or fiction, David Andelman, Guy Deutscher. World Policy Journal, Summer 2015.

SOURCES

CO2 concentrations : CDIAC, http://ornl.gov./trends/CO2

CO2 emissions : World Bank Data Center.

Limits to growth scenario : Graham Turner, *op. cit.*

Epica Dome C data : D. Luthi *et al.*, Nature **453**, 379 (2008).

Carottages Law Dome : Etherige *et al.* (CSIRO), after C.E. Ophard, Virtual Chembook, Elhurst College (2003).

CO2 and temperature correlations (Fig. 3.10) : J.R. Petit *et al.*, Nature **399**, 329 (1999), cité par G. Deutscher, The Entropy Crisis.

Anomalies de température et émissions de CO2 : Rapport du GIEC à l'intention des décideurs, 2013.

Isotope datation : Dome Fuji Ice Core Consortium.

Food and Wood production : Food and Agriculture Organization of the United Nations.

INDEX